Analysis of variance in statistical image processing

Analysis of variance in statistical image processing

LUDWIK KURZ and M. HAFED BENTEFTIFA

Polytechnic University

PUBLISHED BY THE PRESS SYNDICATE OF THE UNIVERSITY OF CAMBRIDGE
The Pitt Building, Trumpington Street, Cambridge CB2 1RP, United Kingdom

CAMBRIDGE UNIVERSITY PRESS
The Edinburgh Building, Cambridge CB2 2RU, United Kingdom
40 West 20th Street, New York, NY 10011-4211, USA
10 Stamford Road, Oakleigh, Melbourne 3166, Australia

First published 1997

Printed in the United States of America

Typeset in Times Roman

Library of Congress Cataloging-in-Publication Data

Kurz, Ludwik.
    Analysis of variance in statistical image processing / Ludwik
Kurz, M. Hafed Benteftifa.
        p.   cm.
    ISBN 0-521-58182-6 (hardback)
    1. Image processing—Statistical methods.   2. Analysis of
variance. I. Benteftifa, M. Hafed.   II. Title.
TA1637.K87   1997   96-22310
621.36′7′01519538—dc20   CIP

A catalog record for this book is available from the British Library

ISBN   0 521 58182 6 hardback

To our families and *Moncef* in particular

CONTENTS

PREFACE

The philosophical approaches pursued in this book can be divided into two groups: imaging problems involving relatively high signal-to-noise ratio (SNR) environments and problems associated with environments in which the images are corrupted by severe, usually non-Gaussian noise. The former class of problems led to the development of numerous approaches involving the so-called experimental design techniques.[2–5] The latter class of problems was addressed using techniques based on partition tests (Kurz,[72] chapter III, Kersten and Kurz[73]). The material in this book is based on experimental design techniques; it represents a graduate course offered by the senior author. The book considers the basic notions involving experimental design techniques. It is hoped that, in addition to being a text for a graduate course, the book will generate interest among imaging engineers and scientists, resulting in further development of algorithms and techniques for solving imaging problems.

The basic problems addressed in the book are line and edge detection, object detection, and image segmentation. The class of test statistics used is mainly based on various forms of the linear model involving analysis of variance (ANOVA) techniques in the framework of experimental designs. Though the statistical model is linear, the actual operations involving imaging data are nonlinear. It has been shown by Scheffe (Chapter10)[2] that statistical tests based on ANOVA are relatively insensitive to the variations in underlying distributions. This is particularly true in processing of imaging data if the corruption of the Gaussian noise background does not exceed ten percent. For higher rates of contamination and for other significant deviations from the Gaussian model, one can robustize the procedure along the lines suggested by Hampel et al.[84] The latter approach is cumbersome and computationally inefficient. A more practical approach to robustizing the ANOVA procedure is to use stochastic approximation algorithms as presented in Chapter 8.

The F-statistic, which is fundamental to the use of techniques based on the ANOVA model, is obtained from the ratio of two chi-square random variables. For the case of testing a null hypothesis against alternatives for a Gaussian noise distribution of unknown and/or variable variance, it is optimum. Namely, the same test maximizes the power of the test (probability of detection) for all alternatives and among all invariant tests with respect to shifting, scaling, and orthogonal transformation of the data resulting in a uniformly most powerful test if the underlying noise is Gaussian (Lehman,[1] p. 67).

Since in processing large images the noise variance or SNR can change significantly over various regions of the image, this situation leads to an automatic thresholding scheme that takes into account the histogram taken from the entire image yielding a constant false alarm rate. To ensure the practicality of the procedures, the image needs to be scanned by a localized mask. The mask must be large to yield enough power of the test, yet it must be sufficiently small to ensure the detection of small features or yield good resolution of image details. From many years of experience with imaging data, the authors know that the mask must consider at least 20 pixels (samples) of observable data. This determines the size of the mask and its geometry, which suggests 5×5 and 3×7 for most applications.

The standard ANOVA theory (independent sampling) is relatively insensitive to the departure from the Gaussian distribution of noise and to the departure from the assumption of equal variance. This insensitivity does not extend to the dependent noise model (Scheffe,[2] p. 364). As a matter of fact, when using modern image processing techniques, the assumption of independent sampling is invalid. Neglecting the statistical dependence of data results in total breakdown of the processing scheme. To bypass this difficulty, the theory of ANOVA is extended to include dependent data. In the latter situation, if the contamination of noise is severe, the robustizing of the procedures using techniques suggested by Hampel et al. is no longer viable. Yet the modifications of the imaging data extraction process based on the stochastic approximation algorithm (Chapter 8) are practicable. Alternatively, one can use procedures based on partition tests.[74, 77]

It is hoped that this book will stimulate a wider acceptance of the methodology of image processing suggested here. Based on many years of experience, the authors are confident that image processing techniques based on the experimental design model will become a significant tool in solving many imaging problems.

The authors are particularly indebted to the senior author's doctoral students. Without their important contributions, this book could not become a reality. The list of references lets the reader become familiar with additional aspects of the approach to imaging problems based on the experimental design model and provides an opportunity to develop additional techniques to solve various problems in the field.

It is understood that the reader is familiar with basic aspects of probability and statistics. The authors decided that including simple problems and exercises would be counterproductive. In teaching the course, specific image processing problems based on the software developed over the years were assigned.

1

Introduction

1.1 Introductory remarks

In considering image processing problems, it is commonly required that certain types of basic patterns be extracted from a noisy and/or complex scene. For digital image processing techniques considered in this book, one generally implies the processing of two-dimensional data that leads to applications in such a broad spectrum of problems as picture processing, medical data interpretation, underwater and earth sounding, trajectory detection, radargram enhancement, etc. If the noise conditions are favorable, involving high signal-to-noise ratios (SNR), and the corrupting noise is Gaussian with independent identically distributed (i.i.d.) samples (pixels), the classical techniques based on the matched filter theory are applicable. On the other hand, even a small deviation from the Gaussian assumption or the variability of SNR in various parts of the image will severely deteriorate the performance of the matched filter. In image formation by certain optical systems, such as infrared sensors and detectors, unfavorable noise conditions will prevail. As a matter of fact, the distribution function of the image pixels contaminated by noise is seldom known in imaging problems.

Another difficulty associated with processing of large two-dimensional images is that the SNR may vary significantly from region to region. In the latter situation the use of a simple thresholded matched filter will yield false alarm rates and probabilities of detection that will vary unpredictably over the scene. In addition, in analyzing a scene one may encounter unwanted background patterns (structured noise) that will further degrade the performance of a linear matched filter. If one uses a histogram of the image data from the entire scene to adjust the threshold, it may reduce the average false alarm rate, yet the probability of detection will vary from region to region, resulting in an unacceptable performance. In addition, the variable threshold approach will also require additional storage and an increase in processing time.

The difficulties associated with the use of matched filters in processing real images led to the development of techniques based on the analysis of variance

(ANOVA) in the framework of experimental designs, which result in simple non-linear statistical operators (masks) easy to implement and adjust to various needs in processing digital images. For the case of testing a null hypothesis against alternatives for Gaussian noise of unknown variance, the F-statistic associated with the ANOVA model is optimum. As a matter of fact, the test involving the F-statistic maximizes the probability of detection for all alternatives and among all invariant tests with respect to shifting, scaling, and orthogonal transformation of the data. Thus, tests based on the F-statistics involving ANOVA are uniformly most powerful (UMP) in Gaussian noise (Lehmann[1], p. 67). Even if the noise variance is unknown and/or variable throughout the image, the probability of detection remains high even for a fixed threshold operation. Based on many years of experience with various imaging data, it was found that a small mask (5×5 or 3×7) is sufficient to obtain excellent results in applying ANOVA techniques to image processing. Also, this type of processing may be adjusted to suppress naturally structural noise.

If the noise deviates from Gaussian, the F-statistic remains robust.[1] It has been the experience of the senior author that contamination of Gaussian noise up to ten percent does not require any adjustment in the processing of imaging data. Hampel[84] indicates that the ANOVA model is sensitive to noise contamination. This may be true for other types of data and for imaging data with larger than ten percent rate of contamination. A procedure suggested by Hampel[84] to robustize the ANOVA model is impractical in imaging problems. If the imaging data are not statistically independent, the Hampel approach breaks down. It was indicated by Box and Anderson[79] that if the kurtosis $K_u = (E(x - \mu)^4/\sigma^4) - 3$ of the Gaussian noise is between ± 1, which corresponds to a peaked probability density function (pdf) with long and thin tails for positive kurtosis and flat pdf that drops off rapidly to zero for negative kurtosis, the F-statistic remains robust. This is also true if the skewness $s = (E(x - \mu)^3/\sigma^3)$ is positive and less than unity (long tail to the right). If the contamination of noise is severe, or if the noise is non-Gaussian, one needs to use a robustized version of the ANOVA model estimators. From a practical point of view, estimators based on stochastic approximation algorithms are useful. An exposition of the stochastic approximation methodology is given in Chapter 8. The chapter may also introduce the reader to the stochastic approximation methodology if he or she is not familiar with it.

An alternate approach to processing of imaging data in non-Gaussian noise involves the application of partition tests. Readers interested in this methodology are referred to Kurz[72] and Kersten and Kurz[73] and numerous other references on the subject.[86, 87] Since standard ANOVA theory (independent sampling) is extremely

---

[1] By robustness, we mean low sensitivity to deviations from a nominal distribution. The concept was first introduced by Box.[78]

sensitive to the dependence of noise samples, a generalization of the model to include statistical dependence is presented in the book. This complicates some of the derivations and expressions, yet the image processing schemes remain simple. The assumption of dependent sampling is realistic if one considers the fast sampling rate in modern data processing systems. If the dependent noise samples are contaminated by non-Gaussian noise, the stochastic approximation methodology may be extended as suggested by Lomp[82] and Kowalski.[83] Alternately, one may use generalized versions of partition tests.[74, 77, 82, 83]

1.2 Decision mechanism in image processing

From a practical point of view the most important aspect of the methodology based on the ANOVA model is the use of small local operators (masks). These masks must be large enough to ensure high probability of detection, yet small enough not to miss some intricate aspects of the image patterns. As indicated before, many years of experience with imaging data suggest 5×5 and 3×7 masks are useful for most applications, though nonrectangular masks may be used in special applications. The specific type of ANOVA design will be determined by specific applications. This statement will be clarified in subsequent chapters. The choice of the masks in conjunction with the use of computers leads to simple calculations, minimal storage for single local masks, and significant reduction in processing time, which in most cases leads to efficient real-time processing. The process of scanning the image by a mask and calculating the test statistic is analogous to a nonlinear time convolution between a signal function and the impulse response of a matched filter. Depending on the type of mask, a number is generated that is compared to a preselected threshold resulting in acceptance or rejection of the point as belonging to the image feature under study, that is, line or edge element. In evaluating the performance of each procedure, subjective and objective measures are used. The subjective evaluation of the procedures is based on numerous simulation results involving known images and various types of noise. In addition, objective evaluations of the various procedures are accomplished using Monte-Carlo simulations to obtain plots of the probability of detection as a function of correlation coefficients for fixed values of false alarm rates. It is interesting to note that all curves of this type are of V-form, with the dip occurring in the midrange of the correlation coefficient. The deterioration of performance in this region results from false patterns being generated by the dependence of noise. It should be noted that even this dip in performance still yields acceptable results.

It is recommended that the reader becomes familiar with the various algorithms by applying the available software (*SIMEX*) to imaging data. The software then may be used to expand the algorithms to include special variations, modifications, and additions of interest to the user.

1.2.1 Organization of the book

The book is presented in nine chapters. After having been introduced to the motivating concepts in this chapter, the reader may proceed to Chapter 2, which presents the basic mathematical notions of the ANOVA model. First, the problem of parameter estimation and linear hypothesis testing is reviewed. This is followed by a detailed presentation of the one- and two-way designs of the ANOVA model. The mathematical developments are related to the physical features of images such as column and row orientations. The two basic designs are then followed by the description of specialized designs such as Latin squares (LS), Græco-Latin squares (GLS), and symmetric balanced incomplete blocks (SBIB). To provide means for testing the validity of the results that are based on multicomparisons, confidence intervals on which to base our decisions are provided. Fundamental to this approach is the concept of contrast functions, which leads naturally to three important multiple comparison techniques: Tukey method, Scheffe method, and Bonferroni method. Because the Scheffe method has an interesting interpretation that if a test based on the F-statistic is rejected for a confidence level α, then a test based on the contrast function will provide means to locate the specific effect for which the effect is rejected, the latter property makes the Scheffe method particularly attractive in image processing applications. In Chapter 2, whenever the proofs are omitted, the reader is referred to Scheffe[2] for proofs involving contrast functions and related topics. It should be noted that unlike the mathematical literature, which treats only the statistically independent case, the theory presented in this chapter is extended to include dependent noise.

The problem of line detection is the basis for the material presented in Chapter 3. Techniques for the detection of lines of various orientations are presented. Specifically ANOVA techniques, starting with the unidirectional detector, which is capable of discriminating only in one direction at a time, are presented. To sharpen the line discrimination capabilities of the procedure, a shape statistic is introduced based on the Scheffe multicomparison test. The process of line detection is then generalized to include a two-way ANOVA model. Next, the GLS designs permit the detection of lines in four directions simultaneously. These algorithms yield high probability of detection in extracting line features at the expense of increase in computational complexity. The GLS design also permits removal of background interference (structural noise). The SBIB design and Youden square design are applied in conjunction with a double-edge transition type algorithm to detect and locate trajectories in noisy scenes. Various simulations as well as Monte-Carlo evaluations of the procedures in this chapter are presented.

The edge detection problem is considered in Chapter 4 (see references [13], [19]). Like the line problem of Chapter 3, the tests for statistical significance are performed in two steps. First, the test statistic is formed to estimate the pertinent parameters for a given ANOVA model. This is followed by a hypothesis test to detect the presence or absence of a particular edge element associated with a preselected direction. Both gray level and texture edges are considered. At first, uni- and bidirectional edge detectors are described followed by multidirectional edge detectors. In areas where edges are corrupted beyond recognition, edge reconstruction algorithms based on directional masks are introduced by using a gradient operation in the context of a GLS mask. The operation of the texture edge detectors is formulated in terms of hypothesis testing of the model parameter variance. Finally, as in the line detection problem, simulations of the various procedures are included.

In Chapter 5 the approaches presented in Chapters 3 and 4 are extended to include radial processing of imaging data. Radial processing results in higher power (improved detectability) in most image processing operations, accompanied by a small cost in increased processing time. In all other aspects the developments in this chapter parallel the material in Chapters 3 and 4.

The important problem of object detection is considered in Chapter 6. The concept of visual equivalence within the framework of standard arrays leads directly to a transformation-based object detection procedure. The procedure is then generalized to include correlated data. Permutation matrices are used to recover the transformation of elements location between original and transformed data spaces. The pertinent test statistics are based on the two-way layout parameterization in modeling the data. Since for large objects inversion of large matrices would be required, an alternate contrast function procedure defined in terms of the effects in the target and background arrays is used. It is applied together with the column and row statistics to test for heterogeneity of the global array. Next, a contrast-based procedure in which the object to be detected is partitioned into background and target arrays is introduced. Elementary contrasts assigned to each array of the partition are then used to generate a contrast function representing a linear combination of these elementary contrasts. The procedure is then generalized involving the use of orthogonal contrasts. This allows for better discrimination among similar objects. In the final part of this chapter a rotation invariant procedure for object detection is introduced. A complex comformal mapping process is used to reduce rotation to translation, which is followed by any of the detection procedures described above. The procedure also allows us to detect the angle between the test and reference objects. Considering that the process is accomplished in the presence of noise, this represents a significant improvement over existing procedures where noise-free or nearly noise-free conditions are assumed.

In many problems involving image data it is necessary to partition the image into various regions to separate features from the background. This partition process is known as image segmentation and is addressed in Chapter 7. The segmentation procedure is based on a split-merge procedure. In general, image elements are merged sequentially to form larger regions. The process is started with a hierarchical decomposition of the image. Using a nested design approach, regions are merged if they satisfy similarity measures defined in terms of an F-test on region effects.

In Chapter 8 certain complementary mathematical and algorithmic concepts are introduced to aid the reader in implementing the procedures presented in previous chapters. In the first part of the chapter the properties of the F-statistic are discussed. This is followed by the description of the Monte-Carlo simulations needed to generate power (probability of detection) for various procedures. Next, an introductory exposition of the stochastic approximation methodology is presented. Robustized versions of the algorithms are also included. This section may be of interest not only for image processing problems but also for many other estimation problems in engineering and science.

Finally, the material on stochastic approximation introduced in Chapter 8 is applied in Chapter 9 to the image restoration problem. Two classes of restoration procedures based on a two-stage approach are introduced. The first stage consists of an edge detector of GLS or Youden square type. The edge detection preprocessor is then followed by a recursive least square estimator described in Chapter 8. If the computing noise is of high variance, a robustized version of the recursive estimator is used. If the corrupting noise is of the "salt and pepper" type, the second stage of the restoration process is based on the "missing value" approach to the restoration process.

2

Statistical linear models

2.1 Introductory remarks

The problem of the application of the linear model in image processing, as developed by Aron and Kurz,[8] involves the interpretation of the experimental data in terms of effects or treatments. Thus, the initial stage is always the selection of the important features, that is, factors to be taken into account and eventually interpreted following the results of any statistical test based on the linear model. The next stage is the introduction of the hypotheses to be tested based on the model that best fits the objectives, the selected factors, and the available data. Finally, the importance to be attached to the eventual results by means of confidence intervals is delineated.

Test statistics based on the theory of the ANOVA within the context of experimental design have been shown to maximize power for all alternatives among all invariant tests with respect to shifting, scaling, and orthogonal transformation of the data.[1] Used in this context, they are generally referred to as UMP.

Several books are devoted to the subject of the ANOVA and it would be rather meaningless to dwell on the theory in this chapter. Instead, we concentrate on the subject of applying ANOVA in image processing problems by providing simple steps to be followed to extract meaningful information from the available data.

We first describe the general model along with the parameter estimation problem that arises in dealing with ANOVA. We then specialize the results to the one-way and two-way layouts. Also included is a detailed analysis of other commonly used designs, namely, Latin squares, Græco-Latin squares, and the balanced incomplete block designs. Next, we review the techniques that are available in the area of multicomparisons among effects. Finally, a detailed analysis of Scheffe's multicomparison technique is presented.

For an easy introduction to the linear model approach the reader may wish to begin with the excellent book by Neter and Wasserman.[36] Besides numerous examples showing the application of ANOVA to various fields, it also contains descriptions of most of the ANOVA models discussed in the literature.

2.2 Linear models

2.2.1 Parameter estimation

The statistical method of analysis of variance is based on the theory of the general linear model. Each collected observation is partitioned into essentially two basic components. The first term is due to causes that are of interest in the study. These causes can be either controlled or measured and represent physical features that are termed "factors" in the experiment. In image processing applications we always assume that the factor effects have already taken place; therefore, any further control during the processing is precluded. The second term in the partition is the random variation or noise component, essentially due to measurement error and any extraneous effect on the experiment.

Data in most cases are collected with a scanning window (mask) with typical size of 5×5. The number of factors that can be assumed depends to a great extent on the objectives to be accomplished. In addition, factors are chosen to represent physical features within the window. There exist two broad classes of models within ANOVA based on the nature of the assumed factors: fixed effects models and random effects models. Before we explain the differences between the models, it is important to keep in mind that a factor in the experiment, which is represented by levels, corresponds to a physical feature that we are interested in detecting by a statistical test on the levels. Hence, it is important to understand the nature of the levels that represent the factor of interest.

When the levels are chosen from a much larger set of levels, the model is referred to as a random effects model. On the other hand, if the levels are specifically selected without being drawn from a population, the model is referred to as a fixed effects model.

The vector space Ω, in which the mean of the observed n-dimensional data vector $\mathbf{y}$ lies, is described by the following model

$$\Omega : \mathbf{y} = \mathbf{X}^T \boldsymbol{\beta} + \mathbf{e} \tag{2.1}$$

where $\boldsymbol{\beta}$ is a p-dimensional vector of unknown parameters to be estimated, $\mathbf{X}^T$ is an $n \times p$ matrix with $p < n$ and is referred to in the statistical literature as the design matrix. The n-dimensional vector $\mathbf{e}$ is the noise vector with the characteristics

$$E(\mathbf{e}) = \mathbf{0} \tag{2.2}$$

and

$$E(\mathbf{ee}^T) = \sigma^2 \mathbf{K}_f \tag{2.3}$$

where σ^2 is the unknown variance and $\mathbf{K}_f$ is the correlation matrix.[1] For the independent data case, $\mathbf{K}_f$ reduces to the identity matrix. The space Ω also represents the parameter space of β, which is transformable to the space of the mean of $\mathbf{y}$ by $\mathbf{X}^T$. Thus,

$$\Omega : \begin{cases} E(\mathbf{y}) = \mathbf{X}^T \beta \\ E((\mathbf{y} - \mathbf{X}^T \beta)(\mathbf{y} - \mathbf{X}^T \beta)^T) = \sigma^2 \mathbf{K}_f \end{cases} \tag{2.4}$$

Central to our discussion is the vector of parameters β that is ordinarily unknown and must be estimated from the available data. An estimate of β can be found using the least squares approach or, in the case of Gaussian noise models, by the maximum likelihood method.

First, we consider the independent data case where the correlation matrix $\mathbf{K}_f$ is the identity matrix. Given Eq. (2.1), the least squares estimate (LSE) of β, which is denoted by $\widehat{\beta}$, is obtained by the well-known normal equation

$$\mathbf{X}\mathbf{X}^T \widehat{\beta} = \mathbf{X}\mathbf{y} \tag{2.5}$$

Here $\widehat{\beta}$ is not unique unless the p column vectors of $\mathbf{X}^T$ are linearly independent or rank $(\mathbf{X}^T) = p$. In case independence is not satisfied, we must place linearly independent side conditions on the parameter vector β in the form

$$\mathbf{B}^T \beta = \mathbf{0} \tag{2.6}$$

The model (2.1) is thus modified to reflect the additional assumption, in which case we have

$$\begin{pmatrix} \mathbf{X}^T \\ \mathbf{B}^T \end{pmatrix} \widehat{\beta} = \begin{pmatrix} \mathbf{y} \\ \mathbf{0} \end{pmatrix} \tag{2.7}$$

The parameter vector estimate can now be calculated and is given by

$$\widehat{\beta} = (\mathbf{X}\mathbf{X}^T + \mathbf{B}\mathbf{B}^T)^{-1}\mathbf{X}\mathbf{y} \tag{2.8}$$

where the rank of $\mathbf{X}^T$ is p and, thus, $\mathbf{X}\mathbf{X}^T$ is invertible.

When the data are dependent, Eq. (2.8) is no longer valid, and we must derive another estimator based on the dependency of the data as expressed by the correlation matrix $\mathbf{K}_f$. The usual approach in similar cases is to proceed with a prewhitening

---

[1] The mathematical literature treats only the statistically independent case.

transformation so that the transformed data is uncorrelated. This approach has been suggested in references [2], [10], [14], [70].

Consider the correlation matrix $\mathbf{K}_f$;[2] we can always find a nonsingular $n \times n$ matrix $\mathbf{P}$ such that the condition $\mathbf{P}^T \mathbf{K}_f \mathbf{P} = \mathbf{I}$ is satisfied. For the purpose of finding the estimate of the effect vector, we do not normally need to find $\mathbf{P}$ but instead we proceed by deriving the expression for the sum of squares using the least squares approach. Consider the vector of orthogonal data obtained by using the transformation $\mathbf{P}$ on the correlated data. We can write

$$\tilde{\mathbf{y}} = \mathbf{P}^T \mathbf{y} \tag{2.9}$$

and the correlation matrix of the new vector of uncorrelated data is

$$E(\tilde{\mathbf{y}}\tilde{\mathbf{y}}^T) = \mathbf{P}^T E(\mathbf{y}\mathbf{y}^T)\mathbf{P} = \mathbf{P}^T \mathbf{K}_f \mathbf{P} = \mathbf{I} \tag{2.10}$$

The resulting error sum of squares for the dependent case is then derived by first considering the alternative for the orthogonal data

$$\Omega : \tilde{\mathbf{y}} \text{ is } N(\mathbf{P}^T \mathbf{X}^T \boldsymbol{\beta}, \sigma^2 \mathbf{I}) \tag{2.11}$$

where the mean is obtained by using Eqs. (2.4) and (2.9). Thus, the error sum of squares is

$$SS_e(\mathbf{y}, \boldsymbol{\beta}) = (\tilde{\mathbf{y}} - \mathbf{P}^T \mathbf{X}^T \boldsymbol{\beta})^T (\tilde{\mathbf{y}} - \mathbf{P}^T \mathbf{X}^T \boldsymbol{\beta}) \tag{2.12}$$

Using Eq. (2.9), we have

$$SS_e(\mathbf{y}, \boldsymbol{\beta}) = (\mathbf{P}^T \mathbf{y} - \mathbf{P}^T \mathbf{X}^T \boldsymbol{\beta})^T (\mathbf{P}^T \mathbf{y} - \mathbf{P}^T \mathbf{X}^T \boldsymbol{\beta}) \tag{2.13}$$

By factoring out the terms in $\mathbf{P}$ and using the fact that $\mathbf{P}\mathbf{P}^T = \mathbf{K}_f^{-1}$, the resulting sum of squares is

$$SS_e(\mathbf{y}, \boldsymbol{\beta}) = (\mathbf{y} - \mathbf{X}^T \boldsymbol{\beta})^T \mathbf{K}_f^{-1} (\mathbf{y} - \mathbf{X}^T \boldsymbol{\beta}) \tag{2.14}$$

The calculation of the effect vector estimate can now be obtained by differentiating $SS_e(\mathbf{y}, \boldsymbol{\beta})$ with respect to $\boldsymbol{\beta}$ using vector differentiation or any other approach that takes into account the specific nature of the correlation matrix.

2.2.2 Linear hypothesis testing

For the general linear model, the linear hypothesis to be tested is generally the test of the means of populations from which samples were taken. In the case of the

---

[2] $\mathbf{K}_f$ is assumed to be a symmetric positive definite matrix.

ANOVA and in the general case, the hypothesis testing refers to the test of the parameter vector $\boldsymbol{\beta}$. The hypothesis–alternative pair is in the form

$$
\begin{aligned}
H_o &: \mathbf{Z}^T \boldsymbol{\beta} = 0 \\
H_a &: \mathbf{Z}^T \boldsymbol{\beta} \neq 0
\end{aligned}
\tag{2.15}
$$

Eq. (2.15) can also be expressed by

$$
\begin{aligned}
H_o &: E(\mathbf{y}) \in \omega \\
H_a &: E(\mathbf{y}) \in \Omega - \omega
\end{aligned}
\tag{2.16}
$$

where ω is a subspace of the mean vector space Ω. A test statistic that is predominantly used in the univariate analysis of variance is

$$
F_a = \frac{(SS_a(\mathbf{y}, \boldsymbol{\beta}) - SS_e(\mathbf{y}, \boldsymbol{\beta}))/n_a}{SS_e(\mathbf{y}, \boldsymbol{\beta})/n_e}
\tag{2.17}
$$

where $SS_a(\mathbf{y}, \boldsymbol{\beta})$ is the sum of squares associated with H_a. Here, $SS_e(\mathbf{y}, \boldsymbol{\beta})$ and $SS_a(\mathbf{y}, \boldsymbol{\beta})$ are chi-square distributed with $n_e = n - r$ and $n_a = q$ degrees of freedom, respectively. From the definition of the F-distribution, it follows that F_a is distributed as $F_{q,n-r,\delta}$ with q and $n - r$ degrees of freedom and noncentrality parameter δ. The decision threshold for the test is defined as the upper α-point of the F-distribution, which is the value whose probability of being exceeded by the random variable F_a is α. The number of degrees of freedom n_e is calculated by the following rule:

n_e = total number of observations − number of independent parameters that were estimated.

2.3 One-way designs

In this section we introduce the basic one-way design, which should provide an easy introduction to the problem of using the linear model in image processing. In this case we assume that only one effect has taken place. What we need to test is whether or not the effect is present in the design.

We consider the case where we have m populations where the sample size for each population is finite and equal to n.[3] The one-way design can be cast within the framework of mean testing in single classification problems. Recalling Eq. (2.1),

---

[3]The assumption is not restrictive in the sense that the model is still valid in the general case of unequal sample sizes.[2]

the model parameterizing the observations is

$$\Omega : \begin{cases} y_{ij} = \mu + \alpha_i + e_{ij} & i = 1, 2, \ldots, m; j = 1, 2, \ldots, n. \\ (e_{ij}) \text{ independent } N(0, \sigma^2 \mathbf{I}) \end{cases} \tag{2.18}$$

where μ and α_i are commonly referred to as the general mean and the effects, respectively. The effects in the one-way model may be assigned to represent row or column effects; the specific choice depends to a great extent on the physical interpretation of the problem at hand. Column effects are best represented by using the notation β_j, whereas α_i denotes row effects.

Thus, with any observation we associate three terms: a term μ that is common to all observations in the sampling unit, a term α_i that is specific to observations aligned in the specific row i, and a random term (e_{ij}) that accounts for errors within the design. The hypothesis is then H_a : all $\alpha_i = 0$, which is tested using the test statistic given in Eq. (2.17). Therefore, we need to derive an expression for the sum of squares under both the alternative and the hypothesis.

Under Ω, the least squares approach leads to choosing estimates of μ and α_i, which minimize

$$SS_e(\mathbf{y}, \boldsymbol{\beta}) = \sum_{i=1}^{m} \sum_{j=1}^{n} (y_{ij} - \mu - \alpha_i)^2 \tag{2.19}$$

Before we proceed any further, note that the design matrix for the one-way layout, shown in Fig. 2.1, is not of full rank. For example, the last column is the sum of the other columns. Therefore, additional side conditions are in order and will be introduced later.

y_{ij}	α_1	α_2	...	α_m	μ
y_{11}	1	0	...	0	1
y_{12}	1	0	...	0	1
.	.	.	...	.	.
y_{1n}	1	0	...	0	1
y_{21}	0	1	...	0	1
.	.	.	...	.	.
y_{2n}	0	1	...	0	1
.	.	.	...	.	.
y_{m1}	0	0	...	1	1
y_{m2}	0	0	...	1	1
.	.	.	...	.	1
y_{mn}	0	0	...	1	1

Figure 2.1. Design matrix $\mathbf{X}^T$.

The values of μ and α_i for which we have a minimum are then found by differentiating Eq. (2.19) with respect to each parameter and equating the results to zero. This will result in a set of two equations that are referred to as the normal equations for the one-way design. By taking the derivative with respect to μ, we obtain

$$\frac{\partial SS_e\,(\mathbf{y}, \boldsymbol{\beta})}{\partial \mu} = 2 \sum_{i=1}^{m} \sum_{j=1}^{n} (y_{ij} - \mu - \alpha_i) = 0 \tag{2.20}$$

and thus the estimate is

$$\widehat{\mu} = \frac{\sum_{i=1}^{m} \sum_{j=1}^{n} y_{ij}}{mn} - \frac{\sum_{i=1}^{m} \alpha_i}{m} \tag{2.21}$$

A common assumption at this stage is $\sum_{i=1}^{m} \alpha_i = 0$. The zero-sum condition on the row effects is generally referred to as the side condition on the effects.[2] Consequently, Eq. (2.21) reduces to

$$\widehat{\mu} = \frac{\sum_{i=1}^{m} \sum_{j=1}^{n} y_{ij}}{mn} = y_{..} \tag{2.22}$$

where the dot represents averaging over the specific index. Using the same approach for the row effects estimates, we have

$$\frac{\partial SS_e\,(\mathbf{y}, \boldsymbol{\beta})}{\partial \alpha_i} = 2 \sum_{j=1}^{n} (y_{ij} - \mu - \alpha_i) = 0 \tag{2.23}$$

and the estimate is then

$$\widehat{\alpha}_i = \frac{\sum_{j=1}^{n} y_{ij}}{n} - \widehat{\mu} = y_{i.} - y_{..} \tag{2.24}$$

The minimum of $SS_e\,(\mathbf{y}, \boldsymbol{\beta})$ is then obtained by replacing the respective estimates in Eqs. (2.22) and (2.24) in Eq. (2.19). Hence,

$$SS_e\,(\mathbf{y}, \boldsymbol{\beta}) = \sum_{i=1}^{m} \sum_{j=1}^{n} (y_{ij} - y_{i.})^2 \tag{2.25}$$

Under H_a all effects are null. In this case it can be shown that the estimate for the general mean is identical to the one given in Eq. (2.22). Therefore, the minimum of the sum of squares under the hypothesis is

$$SS_a\,(\mathbf{y}, \boldsymbol{\beta}) = \sum_{i=1}^{m} \sum_{j=1}^{n} (y_{ij} - y_{..})^2 \tag{2.26}$$

The numerator of the test statistic is obtained by subtracting the sum of squares under the alternative from the sum of squares under the hypothesis; namely $SS_H\,(\mathbf{y}, \boldsymbol{\beta}) = SS_a\,(\mathbf{y}, \boldsymbol{\beta}) - SS_e\,(\mathbf{y}, \boldsymbol{\beta})$.

Thus, using Eqs. (2.25) and (2.26), $SS_H(\mathbf{y}, \beta)$ reduces to

$$SS_H(\mathbf{y}, \beta) = \sum_{i=1}^{m} \sum_{j=1}^{n} (y_{i.} - y_{..})^2 \tag{2.27}$$

Eq. (2.27) measures the spread of the population means $y_{i.}$ from the general mean. Therefore, it is commonly called the "between sum of squares" (BSS) to contrast it with $SS_e(\mathbf{y}, \beta)$, which measures the spread of the observations from each mean within each population and is referred to as the "within sum of squares" (WSS).

There are $m + 1$ parameters with one side condition in the design; therefore, the number of degree of freedom associated with the quadratic in Eq. (2.25) is $m^2 - m$. The number of degrees of freedom associated with Eq. (2.27) is $m - 1$.

Substituting Eqs. (2.25) and (2.27) in Eq. (2.17) and taking care of the respective degrees of freedom, the test statistic for testing the presence of row effects reduces to

$$F_a = \frac{\sum_{i=1}^{m} \sum_{j=1}^{n} (y_{i.} - y_{..})^2 / (m - 1)}{\sum_{i=1}^{m} \sum_{j=1}^{n} (y_{ij} - y_{i.})^2 / (m^2 - m)} \tag{2.28}$$

and the threshold for this design is $F_{\alpha, m-1, m^2-m}$. The hypothesis is rejected whenever the computed value of the test statistic in Eq. (2.28) exceeds the tabulated threshold. At this stage it is important to note that if a test on the presence of vertical effects is desired, it is a simple matter of changing α_i to β_j and carrying out the same analysis.

2.4 Two-way designs

In this section we deal with the case where there is an additional effect in the model. In addition to the global mean, the row effect, and the error term, the effect of column orientation is taken into account by adding a column effect. Thus, observations within a given column j where $j = 1, \ldots, n$ will have a common effect we denote by β_j.

The main advantage in using this additional parameter lies in the fact that two possible tests can now be defined, and therefore the tests of presence of column and row effects can be carried out simultaneously. It is important to note in the following analysis that the size of each population j, $j = 1, 2, \ldots, n$, is equal to m. The case where the size of each population is different from m is more complicated to handle. In application to image processing problems, we usually can control m (m being the size of the scanning window).

The fixed effect model is thus represented by

$$\Omega : \begin{cases} y_{ij} = \mu + \alpha_i + \beta_j + e_{ij} & i = 1, 2, \ldots, m; j = 1, 2, \ldots, n. \\ (e_{ij}) \text{ independent } N(0, \sigma^2 \mathbf{I}) \end{cases} \tag{2.29}$$

In addition, the side conditions on the effects are $\sum_{i=1}^{m} \alpha_i = 0$ and $\sum_{j=1}^{n} \beta_j = 0$.

If we assume that the interactions between effects are negligible,[2] the hypotheses of interest are based on the test of presence of row or column effects, respectively. In terms of α_i and β_j, the hypotheses can be expressed by the following pair

$$H_a : \text{all } \alpha_i = 0$$
$$H_b : \text{all } \beta_j = 0$$
(2.30)

To find the estimates for α_i and β_j, the same approach used in the one-way design is followed. First, the sum of squares under Ω is

$$SS_e(\mathbf{y}, \boldsymbol{\beta}) = \sum_{i=1}^{m} \sum_{j=1}^{n} (y_{ij} - \mu - \alpha_i - \beta_j)^2$$
(2.31)

The normal equations are then obtained through successive differentiation. Namely, for the general mean, we have

$$\frac{\partial SS_e(\mathbf{y}, \boldsymbol{\beta})}{\partial \mu} = 2 \sum_{i=1}^{m} \sum_{j=1}^{n} (y_{ij} - \mu - \alpha_i - \beta_j) = 0$$
(2.32)

Consequently, the mean estimate in terms of the observations and taking into account the side conditions is

$$\hat{\mu} = y_{..}$$
(2.33)

Similarly, for the row effects α_i, we have

$$\frac{\partial SS_e(\mathbf{y}, \boldsymbol{\beta})}{\partial \alpha_i} = 2 \sum_{j=1}^{n} (y_{ij} - \mu - \alpha_i - \beta_j) = 0$$
(2.34)

Expanding the summation and using the side conditions, Eq. (2.34) reduces to

$$\hat{\alpha}_i = y_{i.} - y_{..} \quad i = 1, 2, \ldots, m$$
(2.35)

Finally, for the column effects, we have

$$\frac{\partial SS_e(\mathbf{y}, \boldsymbol{\beta})}{\partial \beta_j} = 2 \sum_{i=1}^{m} (y_{ij} - \mu - \alpha_i - \beta_j) = 0$$
(2.36)

which leads to the estimate

$$\hat{\beta}_j = y_{.j} - y_{..} \quad j = 1, 2, \ldots, n$$
(2.37)

Hence the minimum of $SS_e(\mathbf{y}, \boldsymbol{\beta})$ is obtained by substituting Eqs. (2.33), (2.35), and (2.37) in Eq. (2.31), which yields

$$SS_e(\mathbf{y}, \boldsymbol{\beta}) = \sum_{i=1}^{m} \sum_{j=1}^{n} (y_{ij} - y_{.j} - y_{i.} + y_{..})^2$$
(2.38)

The hypothesis, let us say H_a : all $\alpha_i = 0$, is tested using Eq. (2.17) with the degrees of freedom modified to include the effect of the additional parameter in the design. It is then essential to derive the sum of squares $SS_a(\mathbf{y}, \beta)$ and $SS_b(\mathbf{y}, \beta)$ under H_a and H_b, respectively.

First, when we consider H_a, all the α_i are equal to zero, which implies that we need to retain Eqs. (2.33) and (2.37) as the estimates of μ and β_j. Thus, $SS_a(\mathbf{y}, \beta)$ is given by

$$SS_a(\mathbf{y}, \beta) = \sum_{i=1}^{m}\sum_{j=1}^{n}(y_{ij} - y_{.j})^2 \tag{2.39}$$

Similarly, we have Eqs. (2.33) and (2.35) as the estimates of μ and α_i under H_b, which yields

$$SS_b(\mathbf{y}, \beta) = \sum_{i=1}^{m}\sum_{j=1}^{n}(y_{ij} - y_{i.})^2 \tag{2.40}$$

$$SS_a(\mathbf{y}, \beta) - SS_e(\mathbf{y}, \beta) = \sum_{i=1}^{m}\sum_{j=1}^{n}(y_{i.} - y_{..})^2 \tag{2.41}$$

Similarly,

$$SS_b(\mathbf{y}, \beta) - SS_e(\mathbf{y}, \beta) = \sum_{i=1}^{m}\sum_{j=1}^{n}(y_{.j} - y_{..})^2 \tag{2.42}$$

Note that there are $m + n + 1$ parameters in the design with two side conditions; therefore, the number of degrees of freedom associated with Eq. (2.38) is $mn - m - n + 1$, which can be written as $(m-1)(n-1)$. In addition, $n_a = m - 1$ and $n_b = n - 1$. Consequently, the test statistics for testing H_a and H_b can be expressed as follows

$$F_a = \frac{\sum_{i=1}^{m}\sum_{j=1}^{n}(y_{i.} - y_{..})^2/(m-1)}{\sum_{i=1}^{m}\sum_{j=1}^{n}(y_{ij} - y_{.j} - y_{i.} + y_{..})^2/(m-1)(n-1)} \tag{2.43}$$

and

$$F_b = \frac{\sum_{i=1}^{m}\sum_{j=1}^{n}(y_{.j} - y_{..})^2/(n-1)}{\sum_{i=1}^{m}\sum_{j=1}^{n}(y_{ij} - y_{.j} - y_{i.} + y_{..})^2/(m-1)(n-1)} \tag{2.44}$$

The thresholds associated with the test statistics in Eqs. (2.43) and (2.44) are $F_{\alpha, m-1, (m-1)(n-1)}$ and $F_{\alpha, n-1, (m-1)(n-1)}$, respectively.

2.5 Incomplete designs

Higher-order designs can be defined in a manner similar to the one-way and two-way designs. Consider that there are p effects in the design with the qth effect assuming I_q levels with $q = 1, 2, \ldots, p$. The total number of cells, that is, observations in the one observation per cell case, is $N = I_1.I_2 \ldots I_p$. For example, in a two-way design with the effects assuming m and n levels there are $N = mn$ observations. In practical applications the number of cells is fixed beforehand, usually by choosing the size of the scanning window during the data collection stage. With typical window size of 5×5, the constraint on the maximum number of effects is such that we are in general unable to define more effects than we hope to include in the design. We resort in this case to what is referred to as the incomplete p-way design. As a result, some of the cells will have no observations. This is to be contrasted with the complete layout where all cells have at least one observation. Therefore, in the incomplete design the main limitation on the maximum number of effects that can be used in the parameterization given the finite number of observations is eliminated.

The main advantage of incomplete designs over lower-order designs lies in the fact that more effects can be studied for the same number of observations or cells, hence allowing the modeling of more structure within the design.

In this section we introduce three typical layouts that can be cast within the framework of incomplete designs. They are the Latin square, Græco-Latin square, and incomplete block design.

2.5.1 Latin square design

The Latin square design is an incomplete three-way layout in which all three factors are at the same number m of levels. The observations are taken according to some fixed pattern on only m^2 out of the m^3 possible treatment combinations. Fig. 2.2 shows a typical layout used for modeling diagonal effects oriented in the $135°$ direction, this in addition to the usual vertical and horizontal effects. The additional factor in the design is referenced by the Latin letters. The layout is such that each letter appears only once in each row and also in each column.

Consider the set of indexes $S = \{i, j, k\}$ that spans the m^2 possible values. The model for the observations is the following

$$\Omega : \begin{cases} y_{ijk} = \mu + \alpha_i + \beta_j + \tau_k + e_{ijk} & (i, j, k) \in S \\ (e_{ijk}) \text{ independent } N(0, \sigma^2 \mathbf{I}) \\ \sum_i \alpha_i = 0; \sum_{j=1}^n \beta_j = 0; \sum_k \tau_k = 0. \end{cases} \qquad (2.45)$$

The present model does not reflect an interpretation of each effect in terms of physical features. This is generally done by selecting the layout pattern. If at the onset we assign α_i and β_j to represent the horizontal and vertical effects, then the effect τ_k could be freely chosen to represent the additional feature of interest. For example, the layout shown in Fig. 2.3 can be used to model effects oriented in the $45°$ direction.

The hypotheses of interest for the Latin square design in this case are

$$H_a : \text{all } \alpha_i = 0$$
$$H_b : \text{all } \beta_j = 0 \tag{2.46}$$
$$H_c : \text{all } \tau_k = 0$$

Factor I	Factor II				
	β_1	β_2	β_3	β_4	β_5
α_1	A	B	C	D	E
α_2	E	A	B	C	D
α_3	D	E	A	B	C
α_4	C	D	E	A	B
α_5	B	C	D	E	A

Figure 2.2. 5×5 Latin square design. Diagonal features are oriented in $135°$ direction.

Factor I	Factor II				
	β_1	β_2	β_3	β_4	β_5
α_1	A	B	C	D	E
α_2	B	C	D	E	A
α_3	C	D	E	A	B
α_4	D	E	A	B	C
α_5	E	A	B	C	D

Figure 2.3. 5×5 Latin square design. Diagonal features are oriented in $45°$ direction.

Under each hypothesis, we must derive the test statistic that is similar in form to Eq. (2.17).

As usual, we first consider the sum of squares under Ω, which is given by

$$SS_e(\mathbf{y}, \boldsymbol{\beta}) = \sum_{(i,j,k)\in S} (y_{ijk} - \mu - \alpha_i - \beta_j - \tau_k)^2 \tag{2.47}$$

Consider the determination of the estimate of μ. We have in this case

$$\frac{\partial SS_e(\mathbf{y}, \boldsymbol{\beta})}{\partial \mu} = 2 \sum_{(i,j,k)\in S} (y_{ijk} - \mu - \alpha_i - \beta_j - \tau_k) = 0 \tag{2.48}$$

Expanding the summation, we obtain

$$\sum_{(i,j,k)\in S} \mu = \sum_{(i,j,k)\in S} y_{ijk} - \sum_{(i,j,k)\in S} \alpha_i - \sum_{(i,j,k)\in S} \beta_j - \sum_{(i,j,k)\in S} \tau_k \qquad (2.49)$$

Since the set S of indexes involves only m^2 values of the m^3 possible ones, the following side conditions are generally assumed for the Latin square

$$\sum_{(i,j,k)\in S} \alpha_i = m \sum_i \alpha_i = 0 \qquad (2.50)$$

$$\sum_{(i,j,k)\in S} \beta_j = m \sum_{j=1}^n \beta_j = 0 \qquad (2.51)$$

$$\sum_{(i,j,k)\in S} \tau_k = m \sum_k \tau_k = 0 \qquad (2.52)$$

Hence, Eq. (2.49) reduces to

$$\widehat{\mu} = \frac{\sum_{(i,j,k)\in S} y_{ijk}}{m^2} = y_{...} \qquad (2.53)$$

Differentiating $SS_e\,(\mathbf{y}, \boldsymbol{\beta})$ with respect to α_i, we obtain

$$\frac{\partial SS_e\,(\mathbf{y}, \boldsymbol{\beta})}{\partial \alpha_i} = 2 \sum_{(j,k)\in S} (y_{ijk} - \mu - \alpha_i - \beta_j - \tau_k) = 0 \qquad (2.54)$$

Using the side conditions and expanding Eq. (2.54), we obtain the estimate of the row effect α_i

$$\widehat{\alpha}_i = \frac{\sum_{(j,k)\in S} y_{ijk}}{m^2} - \widehat{\mu} = y_{i..} - y_{...} \quad i = 1, 2, \dots, m \qquad (2.55)$$

A similar procedure leads to the estimates of β_j and τ_k

$$\widehat{\beta}_j = y_{.j.} - y_{...} \quad j = 1, 2, \dots, m \qquad (2.56)$$

and

$$\widehat{\tau}_k = y_{..k} - y_{...} \quad k = 1, 2, \dots, m \qquad (2.57)$$

Hence, the minimum of $SS_e\,(\mathbf{y}, \boldsymbol{\beta})$ is

$$SS_e\,(\mathbf{y}, \boldsymbol{\beta}) = \sum_{(i,j,k)\in S} (y_{ijk} - y_{i..} - y_{.j.} + y_{..k} + 2y_{...})^2 \qquad (2.58)$$

Now, it remains to find the sum of squares $SS_a\,(\mathbf{y}, \boldsymbol{\beta})$, $SS_b\,(\mathbf{y}, \boldsymbol{\beta})$, and $SS_c\,(\mathbf{y}, \boldsymbol{\beta})$ corresponding to hypotheses H_a, H_b, and H_c, respectively. Using the fact that the estimates of the effects, excluding the effect for which the hypothesis is considered, are identical to the estimates under the alternative, it is then easy to show that

$$SS_a\left(\mathbf{y}, \boldsymbol{\beta}\right) = \sum_{(i,j,k)\in S} \left(y_{ijk} - y_{.j.} + y_{..k} + y_{...}\right)^2 \tag{2.59}$$

$$SS_b\left(\mathbf{y}, \boldsymbol{\beta}\right) = \sum_{(i,j,k)\in S} \left(y_{ijk} - y_{i..} + y_{..k} + y_{...}\right)^2 \tag{2.60}$$

and

$$SS_c\left(\mathbf{y}, \boldsymbol{\beta}\right) = \sum_{(i,j,k)\in S} \left(y_{ijk} - y_{i..} + y_{.j.} + y_{...}\right)^2 \tag{2.61}$$

By subtracting $SS_e\left(\mathbf{y}, \boldsymbol{\beta}\right)$ from the respective sum of squares in Eqs. (2.59), (2.60), and (2.61), we obtain

$$SS_a\left(\mathbf{y}, \boldsymbol{\beta}\right) - SS_e\left(\mathbf{y}, \boldsymbol{\beta}\right) = \sum_{(i,j,k)\in S} \left(y_{i..} - y_{...}\right)^2 \tag{2.62}$$

$$SS_b\left(\mathbf{y}, \boldsymbol{\beta}\right) - SS_e\left(\mathbf{y}, \boldsymbol{\beta}\right) = \sum_{(i,j,k)\in S} \left(y_{.j.} - y_{...}\right)^2 \tag{2.63}$$

and finally,

$$SS_c\left(\mathbf{y}, \boldsymbol{\beta}\right) - SS_e\left(\mathbf{y}, \boldsymbol{\beta}\right) = \sum_{(i,j,k)\in S} \left(y_{..k} - y_{...}\right)^2 \tag{2.64}$$

The total number of observations in the design is m^2. Under the Ω assumptions there are $3m+1$ parameters with three side conditions. Thus, the degree of freedom n_e corresponding to Eq. (2.58) is equal to $(m-1)(m-2)$.

Since the numbers of levels for each effect are all equal, then $n_a = n_b = n_c = (m-1)$.

As a result, the test statistics for testing the hypotheses H_a : all $\alpha_i = 0, H_b$: all $\beta_j = 0$ and H_c : all $\tau_k = 0$ are given by

$$F_a = \frac{\sum_{(i,j,k)\in S}\left(y_{i..} - y_{...}\right)^2}{\sum_{(i,j,k)\in S}(y_{ijk} - y_{i..} - y_{.j.} - y_{..k} + 2y_{...})^2/(m-2)} \tag{2.65}$$

$$F_b = \frac{\sum_{(i,j,k)\in S}\left(y_{.j.} - y_{...}\right)^2}{\sum_{(i,j,k)\in S}(y_{ijk} - y_{i..} - y_{.j.} - y_{..k} + 2y_{...})^2/(m-2)} \tag{2.66}$$

and

$$F_c = \frac{\sum_{(i,j,k)\in S}\left(y_{..k} - y_{...}\right)^2}{\sum_{(i,j,k)\in S}(y_{ijk} - y_{i..} - y_{.j.} - y_{..k} + 2y_{...})^2/(m-2)} \tag{2.67}$$

The thresholds are identically equal to $F_{\alpha,(m-1),(m-1)(m-2)}$.

2.5.2 Græco-Latin square design

In practical applications it is often advantageous to include an additional fourth
effect in the model that will enable the simultaneous detection of, for example,
diagonal effects oriented in the 45° and 135° directions.

The configuration where we have two orthogonal Latin squares is commonly
referred to as the Græco-Latin square. It is a particular class of the four-way
models in ANOVA. Each observation in the layout is the sum of four parameters,
each one representing a particular treatment and a noise component. The general
model is described by

$$\Omega : \begin{cases} y_{ijkl} = \mu + \alpha_i + \beta_j + \tau_k + \delta_l + e_{ijkl} & (i, j, k, l) \in S \\ (e_{ijk}) \text{ independent } N(0, \sigma^2 \mathbf{I}) \\ \sum_i \alpha_i = 0; \sum_{j=1}^n \beta_j = 0; \sum_k \tau_k = 0; \sum_l \delta_l = 0. \end{cases} \tag{2.68}$$

where the (e_{ijkl}) are normal with zero mean and common variance σ^2. As compared
to the Latin square design, the additional effect is δ_l.

Factor I	Factor II				
	β_1	β_2	β_3	β_4	β_5
α_1	$A : \alpha$	$B : \beta$	$C : \gamma$	$D : \delta$	$E : \epsilon$
α_2	$E : \beta$	$A : \gamma$	$B : \delta$	$C : \epsilon$	$D : \alpha$
α_3	$D : \gamma$	$E : \delta$	$A : \epsilon$	$B : \alpha$	$C : \beta$
α_4	$C : \delta$	$D : \epsilon$	$E : \alpha$	$A : \beta$	$B : \gamma$
α_5	$B : \epsilon$	$C : \alpha$	$D : \beta$	$E : \gamma$	$A : \delta$

Figure 2.4. 5×5 Græco-Latin square design.

S is the set of the m^2 possible observations in the design. Note that the noise
factor can be characterized by two indexes only. Thus, in subsequent derivations
we will use the pair (i, j) to reference each noise and data component in the model.
From Fig. 2.4 we see that the Græco-Latin square is obtained by juxtaposing two
Latin squares. Each one is represented by either the Latin or Greek letter, giving
rise to the name of the design.

The four treatment groups in the design are the rows, the columns, and the Greek
and Latin letters. The Greek and Latin letters, which are at the intersection of
a specific row and column, are arranged in a systematic manner, parallel to the
diagonals. Every treatment is coupled with other treatments exactly once in the
case of this specific Græco-Latin square. Fig. 2.5 shows the particular orientation
of the effects under each Latin design.

The four hypotheses to be tested are

$$H_a : \text{all } \alpha_i = 0$$
$$H_b : \text{all } \beta_j = 0$$
$$H_c : \text{all } \tau_k = 0$$
$$H_d : \text{all } \delta_l = 0$$

(2.69)

Factor I	Factor II				
	β_1	β_2	β_3	β_4	β_5
α_1	A	B	C	D	E
α_2	E	A	B	C	D
α_3	D	E	A	B	C
α_4	C	D	E	A	B
α_5	B	C	D	E	A

Figure 2.5. Latin square 135°.

Factor I	Factor II				
	β_1	β_2	β_3	β_4	β_5
α_1	α	β	γ	δ	ϵ
α_2	β	γ	δ	ϵ	α
α_3	γ	δ	ϵ	α	β
α_4	δ	ϵ	α	β	γ
α_5	ϵ	α	β	γ	δ

Figure 2.6. Latin square 45°.

The sum of squares to be minimized under Ω is

$$SS_e(\mathbf{y}, \boldsymbol{\beta}) = \sum_{(i,j,k,l) \in S} (y_{ij} - \mu - \alpha_i - \beta_j - \tau_k - \delta_l)^2$$

(2.70)

The normal equations are obtained in the same manner as in Section 2.5.1. It is straightforward to show that the estimates for the respective effects are given by

$$\begin{aligned}
\widehat{\mu} &= y_{....} \\
\widehat{\alpha}_i &= y_{i...} - y_{....} & i &= 1, 2, \ldots, m \\
\widehat{\beta}_j &= y_{.j..} - y_{....} & j &= 1, 2, \ldots, m \\
\widehat{\tau}_k &= y_{..k.} - y_{....} & k &= 1, 2, \ldots, m \\
\widehat{\delta}_l &= y_{...l} - y_{....} & l &= 1, 2, \ldots, m
\end{aligned}$$

(2.71)

The minimum of $SS_e(\mathbf{y}, \boldsymbol{\beta})$ is then obtained by replacing Eq. (2.71) in (2.70), which yields

$$SS_e(\mathbf{y}, \boldsymbol{\beta}) = \sum_{(i,j,k,l)\in S} (y_{ij} - y_{i...} - y_{.j..} - y_{..k.} - y_{...l} + 3y_{....})^2 \qquad (2.72)$$

Under Ω, all effects are present in the design, therefore there are $4m+1$ parameters that are subject to *four* side conditions. Thus, the number of degrees of freedom associated with the quadratic form $SS_e(\mathbf{y}, \boldsymbol{\beta})$ is $n - r$ where $n = m^2$ and $r = 4m + 1 - 4 = 4m - 3$. Hence, $n_e = m^2 - 4m + 3 = (m-1)(m-3)$.

Now that we have the denominator of the F-statistic for the test of the hypotheses, we need to derive the sum of squares under each hypothesis.

Beside the effect for which the hypothesis is considered, all the remaining effects will have the same estimates as under the alternative.

Thus, one can show that under H_a

$$
\begin{aligned}
\widehat{\mu} &= y_{....} & & \\
\widehat{\beta}_j &= y_{.j..} - y_{....} & j &= 1, 2, \ldots, m \\
\widehat{\tau}_k &= y_{..k.} - y_{....} & k &= 1, 2, \ldots, m \\
\widehat{\delta}_l &= y_{...l} - y_{....} & l &= 1, 2, \ldots, m
\end{aligned}
\qquad (2.73)
$$

and the minimum of $SS_a(\mathbf{y}, \boldsymbol{\beta})$ is

$$SS_a(\mathbf{y}, \boldsymbol{\beta}) = \sum_{(i,j,k,l)\in S} (y_{ij} - y_{.j..} - y_{..k.} - y_{...l} + 2y_{....})^2 \qquad (2.74)$$

Finally, the numerator of the F-test is

$$SS_a(\mathbf{y}, \boldsymbol{\beta}) - SS_e(\mathbf{y}, \boldsymbol{\beta}) = \sum_{(i,j,k,l)\in S} (y_{i...} - y_{....})^2 \qquad (2.75)$$

and the number of degree of freedom associated with $SS_a(\mathbf{y}, \boldsymbol{\beta})$ is $m - 1$.

Hence, the F-statistic for testing H_a is obtained by forming the ratio of Eqs. (2.75) and (2.72) with the proper degree of freedom assigned to each quadratic. Thus, we have

$$F_a = \frac{(m-3)\sum_{(i,j,k,l)\in S}(y_{i...} - y_{....})^2}{\sum_{(i,j,k,l)\in S}(y_{ij} - y_{i...} - y_{.j..} - y_{..k.} - y_{...l} + 3y_{....})^2} \qquad (2.76)$$

A similar approach may be used for the determination of the statistics under H_b, H_c, and H_d. In which case we obtain

$$F_b = \frac{(m-3)\sum_{(i,j,k,l)\in S}(y_{.j..} - y_{....})^2}{\sum_{(i,j,k,l)\in S}(y_{ij} - y_{i...} - y_{.j..} - y_{..k.} - y_{...l} + 3y_{....})^2} \qquad (2.77)$$

$$F_c = \frac{(m-3)\sum_{(i,j,k,l)\in S}(y_{..k.} - y_{....})^2}{\sum_{(i,j,k,l)\in S}(y_{ij} - y_{i...} - y_{.j..} - y_{..k.} - y_{...l} + 3y_{....})^2} \qquad (2.78)$$

and

$$F_d = \frac{(m-3)\sum_{(i,j,k,l)\in S}(y_{...l} - y_{....})^2}{\sum_{(i,j,k,l)\in S}(y_{ij} - y_{i...} - y_{.j..} - y_{..k.} - y_{...l} + 3y_{....})^2} \qquad (2.79)$$

The threshold for each test is $F_{\alpha,m-1,(m-1)(m-3)}$.

2.5.3 Incomplete block design

Under the two-way layout with fixed effects, we considered a layout with m row and n column effects, respectively. The design lends itself relatively well to the test of presence of row or column effects without any further assumptions about the collected experimental data. However, in some cases it is desirable to group the data, considered in terms of experimental units, into homogeneous groups or blocks. In this case, and retaining the philosophy governing the two-way layout, the effects are now defined in terms of blocks and treatments within blocks. This particular approach often allows more refined and precise comparisons of the treatments than in the standard two-way design.

Once the number of blocks within the design is decided, the treatments are usually assigned in a random manner within the blocks. The purpose of randomization is two-fold. First, it allows the elimination of external, uncontrolled effects not under consideration, that would otherwise introduce bias in the study. This is done regardless of whether they are present or not. In other words, randomization is a precautionary measure. Second, it is possible to draw sound conclusions from a test in which randomization was incorporated in the initial stages of the study.

Blocks	I	II	III
	B	C	A
	C	D	B
	A	B	D
	D	A	C

Figure 2.7. Complete randomized block design. Four treatments in three blocks.

Randomization is achieved by using permutation tables such as those given in reference [4]. Often, there are additional constraints on the way to achieve the necessary randomization, usually based on such considerations as whether treatments should appear in one or more blocks, etc. Therefore, it is suggested that one must understand what type of analysis on the treatments and blocks is being sought before proceeding with the randomization of the data. The complete block design analysis is often impossible to carry out for several reasons. The main reason is that the sizes of the experimental units are often small compared to the total number of treatments and blocks being considered. As a result, we turn to the incomplete block design in which the number of treatments appearing in any one block is less than the maximum number of treatments in the design.

An important design is the balanced incomplete block (BIB) design: All blocks are of the same size, each treatment appears the same number of times in all blocks and, in addition, each pair of treatment levels appears together in the same block the same number of times in the design.

Blocks	I	II	III	IV	V	VI	VII
	A	F	B	C	E	D	G
	C	A	D	G	B	F	E
	E	G	A	B	F	C	D

Figure 2.8. Balanced incomplete block design. Seven treatments in blocks of three.

The additive model for the BIB design is similar in form to the two-way model with the row and column effects being replaced by block and treatment effects. The observations, in the case of fixed block effects without any interaction between blocks and treatments, are parameterized by

$$\Omega : \begin{cases} y_{ij} = \mu + \alpha_i + \beta_j + e_{ij} & i = 1, 2, \ldots, t; j = 1, 2, \ldots, b. \\ (e_{ij}) \text{ independent } N(0, \sigma^2 \mathbf{I}) \end{cases} \tag{2.80}$$

where μ is the general mean, α_i is the treatment effect subject to the condition $\sum_{i=1}^{t} \alpha_i = 0$, and β_j is the block effects subject to $\sum_{j=1}^{b} \beta_j = 0$.

The hypotheses of interest are the following

$$H_a : \text{all } \alpha_i = 0, \text{ that is, no treatments effects are present.}$$
$$H_b : \text{all } \beta_j = 0, \text{ that is, no block effects are present.} \tag{2.81}$$

Consider the following parameters for the BIB design:
- b: number of blocks in the design
- t: number of treatments in the design
- k: number of treatments in a block.

The error sum of squares under Ω is then

$$SS_e (\mathbf{y}, \boldsymbol{\beta}) = \sum_{i=1}^{t} \sum_{j=1}^{b} (y_{ij} - \mu - \alpha_i - \beta_j)^2 n_{ij} \tag{2.82}$$

where n_{ij} is an indicator that relates the treatment–blocks presence in the design, constrained to b, t, k, and n by the relation

$$\sum_{i=1}^{t}\sum_{j=1}^{b} n_{ij} = bk = n \tag{2.83}$$

The same methodology, that is, the least squares approach, is used to derive the estimates of μ, α_i, and β_j. First, we have

$$\frac{\partial SS_e(\mathbf{y}, \boldsymbol{\beta})}{\partial \mu} = 2\sum_{i=1}^{t}\sum_{j=1}^{b}(y_{ij} - \mu - \alpha_i - \beta_j)n_{ij} = 0 \tag{2.84}$$

By expanding the summation, we obtain

$$\sum_{i=1}^{t}\sum_{j=1}^{b} y_{ij}n_{ij} = \mu\sum_{i=1}^{t}\sum_{j=1}^{b} n_{ij} + \sum_{i=1}^{t}\sum_{j=1}^{b}\alpha_i n_{ij} + \sum_{i=1}^{t}\sum_{j=1}^{b}\beta_j n_{ij} \tag{2.85}$$

Define ρ as the number of replicates of each treatment. In the remaining part of this section, we consider the symmetric BIB (SBIB) design in which the number of treatments t is equal to the number of blocks b with the restriction $bk = t\rho$. That implies that ρ is equal to the number of treatments in a block, that is, $\rho = k$.

Now, consider the second term on the right-hand side of Eq. (2.85). Note that $\sum_{j=1}^{b} n_{ij}$ is the replicate of a specific treatment in the design that is actually ρ. Hence, we can write

$$\sum_{i=1}^{t}\sum_{j=1}^{b}\alpha_i n_{ij} = \rho\sum_{i=1}^{t}\alpha_i \tag{2.86}$$

Similarly, consider the last term on the right-hand side of Eq. (2.85) where $\sum_{i=1}^{t} n_{ij}$ is the number of treatment in a block. Thus, we can write

$$\sum_{i=1}^{t}\sum_{j=1}^{b}\beta_j n_{ij} = k\sum_{j=1}^{b}\beta_j \tag{2.87}$$

Finally, substituting Eqs. (2.86) and (2.87) in Eq. (2.85), we obtain

$$\sum_{i=1}^{t}\sum_{j=1}^{b} y_{ij}n_{ij} = n\widehat{\mu} + \rho\sum_{i=1}^{t}\widehat{\alpha}_i + k\sum_{j=1}^{b}\widehat{\beta}_j \tag{2.88}$$

where use has been made of the hat notation to denote the estimate of the effect instead of the actual value. Differentiating with respect to the treatment effects α_i, we obtain

$$\frac{\partial SS_e(\mathbf{y}, \boldsymbol{\beta})}{\partial \alpha_i} = 2\sum_{j=1}^{b}(y_{ij} - \mu - \alpha_i - \beta_j)n_{ij} = 0 \tag{2.89}$$

By expanding the summation and using the same argument involving the indicator n_{ij}, Eq. (2.89) reduces to

$$\sum_{j=1}^{b} y_{ij}n_{ij} = \rho\widehat{\mu} + \rho\widehat{\alpha}_i + \sum_{j=1}^{b} \widehat{\beta}_j n_{ij} \tag{2.90}$$

Finally, by taking the derivative with respect to the β_j, we obtain the relation

$$\sum_{i=1}^{t} y_{ij}n_{ij} = k\widehat{\mu} + k\widehat{\beta}_j + \sum_{i=1}^{t} \widehat{\alpha}_i n_{ij} \tag{2.91}$$

Let

$$\begin{aligned} G &= \sum_{i=1}^{t}\sum_{j=1}^{b} y_{ij}n_{ij} \\ T_i &= \sum_{j=1}^{b} y_{ij}n_{ij} \\ B_j &= \sum_{i=1}^{t} y_{ij}n_{ij} \end{aligned} \tag{2.92}$$

Using the side conditions for the treatment and block effects, we obtain from Eq. (2.91) the estimate of μ

$$\widehat{\mu} = \frac{G}{n} \tag{2.93}$$

Combining Eqs. (2.90) and (2.91) to eliminate the block effects, we obtain

$$T_i = \rho\widehat{\mu} + \rho\widehat{\alpha}_i + \sum_{j=1}^{b} \frac{[B_j - k\widehat{\mu} - \sum_{i=1}^{t}\widehat{\alpha}_i n_{ij}]}{k} n_{ij} \tag{2.94}$$

which can be written as

$$kT_i = k\rho\widehat{\mu} + k\rho\widehat{\alpha}_i + \sum_{j=1}^{b} B_j n_{ij} - k\widehat{\mu}\sum_{j=1}^{b} n_{ij} - \sum_{l=1}^{t}\sum_{j=1}^{b} \widehat{\alpha}_l n_{ij}n_{lj} \tag{2.95}$$

The last term in Eq. (2.95) is expanded into two terms with respect to the index l. Thus, we have

$$\sum_{l=1}^{t}\sum_{j=1}^{b} \widehat{\alpha}_l n_{ij}n_{lj} = \sum_{\substack{l=1\\l\neq i}}^{t}\sum_{j=1}^{b} \widehat{\alpha}_l n_{ij}n_{lj} + \sum_{\substack{j=1\\l=i}}^{b} \widehat{\alpha}_l n_{lj}^2 \tag{2.96}$$

To simplify Eq. (2.96), we use the following relation

$$\sum_{\substack{l=1\\l\neq i}}^{t}\sum_{j=1}^{b} \widehat{\alpha}_l n_{ij}n_{lj} = \sum_{\substack{l=1\\l\neq i}}^{b} \alpha_l(n_{ij}n_{lj}) = \lambda\sum_{\substack{l=1\\l\neq i}}^{t} \alpha_l = \lambda\sum_{l=1}^{t} \alpha_l - \lambda\alpha_i \tag{2.97}$$

where λ is defined as the number of blocks in which two given treatments appear together. Since n_{ij} can be only 1 or 0, we have $n_{ij}^2 = n_{ij}$, hence

$$\sum_{j=1}^{b} n_{ij}^2 = \sum_{j=1}^{b} n_{ij} = \rho \tag{2.98}$$

Finally, by using the side conditions on the treatment effect, Eq. (2.96) reduces to

$$kT_i - \sum_{j=1}^{b} B_j n_{ij} = \widehat{\alpha}_i \left[\lambda + (k-1)\rho\right] \tag{2.99}$$

For the SBIB design, $\lambda = \frac{(k-1)\rho}{(t-1)}$.
Hence from Eq. (2.99), the estimate of α_i is given by

$$\widehat{\alpha}_i = \frac{k}{\lambda t} T_i - \frac{1}{\lambda t} \sum_{j=1}^{b} B_j n_{ij} \quad i = 1, 2, \ldots, t \tag{2.100}$$

By replacing α_i and μ in either Eq. (2.90) or Eq. (2.91) and using some algebraic manipulations, we obtain the estimate of β_j

$$\widehat{\beta}_j = \frac{\rho}{\lambda t} B_j - \frac{1}{\lambda t} \sum_{i=1}^{t} T_i n_{ij} \quad j = 1, 2, \ldots, b \tag{2.101}$$

The error sum of squares under the alternative can now be evaluated by replacing the estimates given in Eqs. (2.93), (2.100), and (2.101) in Eq. (2.82). Hence,

$$SS_e(\mathbf{y}, \boldsymbol{\beta}) = \sum_{i=1}^{t} \sum_{j=1}^{b} \left(y_{ij} - \frac{G}{n} - T_i \frac{k}{\lambda t} + \sum_{j=1}^{b} B_j n_{ij} - B_j \frac{\rho}{\lambda t} + \sum_{i=1}^{t} T_i n_{ij} \right)^2 n_{ij} \tag{2.102}$$

The sum of squares under H_a is obtained by deriving the estimates of μ and the block effects β_j given that all the treatment effects α_i are not present. Note that the least squares estimates under the hypotheses are different from the estimates under the alternatives. This is to be contrasted with previous designs such as the two-way design, where the analysis under the hypotheses is directly related to the analysis under the alternative. Eliminating the terms involving α_i in Eq. (2.82), we obtain

$$SS_a(\mathbf{y}, \boldsymbol{\beta}) = \sum_{i=1}^{t} \sum_{j=1}^{b} (y_{ij} - \mu - \beta_j)^2 n_{ij} \tag{2.103}$$

which leads to the conclusion that as far as the estimate of μ is concerned, it is the same as under Ω. Now, by using Eq. (2.91) and deleting the terms involving α_i, we obtain

$$\widehat{\beta}_j = \frac{1}{k} B_j - \widehat{\mu} \quad j = 1, 2, \ldots, b \tag{2.104}$$

The error sum of squares under H_a is

$$SS_a(\mathbf{y}, \boldsymbol{\beta}) = \sum_{i=1}^{t} \sum_{j=1}^{b} \left(y_{ij} - \frac{B_j}{k} \right)^2 n_{ij} \tag{2.105}$$

Using a similar argument in the case of hypothesis H_b, we obtain

$$SS_b(\mathbf{y}, \boldsymbol{\beta}) = \sum_{i=1}^{t} \sum_{j=1}^{b} \left(y_{ij} - \frac{T_i}{\rho} \right)^2 n_{ij} \tag{2.106}$$

There are bt observations in the design with b and t block and treatment effects. With the side conditions, the degree of freedom associated with $SS_e(\mathbf{y}, \boldsymbol{\beta})$ is $n_e = bt - b - t + 1$ while $n_a = t - 1$ and $n_b = b - 1$, respectively.

Consequently, the test statistics for the test of treatment and block effects are

$$F_t = \frac{(SS_a(\mathbf{y}, \boldsymbol{\beta}) - SS_e(\mathbf{y}, \boldsymbol{\beta}))/(t-1)}{SS_e(\mathbf{y}, \boldsymbol{\beta})/(bk - b - t + 1)} \tag{2.107}$$

and

$$F_b = \frac{(SS_b(\mathbf{y}, \boldsymbol{\beta}) - SS_e(\mathbf{y}, \boldsymbol{\beta}))/(b-1)}{SS_e(\mathbf{y}, \boldsymbol{\beta})/(bk - b - t + 1)} \tag{2.108}$$

2.6 Contrast functions

The test of the hypothesis, let us say H_a, in the one-way design, is based mainly on the comparison of the test statistic (Eq. (2.17)) against a tabulated threshold for a given confidence level. In the case where the test is rejected, there is no further information about the specific effect for which the hypothesis is rejected.

Throughout this book we will be interested in comparisons among effects, either vertical, horizontal, or diagonal, depending on the specific problem at hand. Therefore, it is necessary to provide some means to test the validity of the results of our comparisons; namely, providing confidence intervals on which to base our decisions. This approach has seen rather extensive application in statistical applications and is well established.[6]

Of interest to us are three tests that are commonly used in problems involving comparisons among effects. Before we embark on the definition of each comparison technique, we define some useful terms.

Consider that in a given design, we have p effects $\beta_1, \ldots, \beta_p$, then

- A pairwise function among the effects is of the form $\beta_i - \beta_j$ where $i = 1, 2, \ldots, p; j = 1, \ldots, p$ and $i \neq j$.

- A contrast function among the effects is of the form $\sum_{j=1}^{p} c_j \beta_j$ where $\sum_{j=1}^{p} c_j = 0$.
- A linear combination of the effects is of the form $\sum_{j=1}^{p} c_j \beta_j$.

Notice that the pairwise as well as the contrast functions are special cases of the linear combination functions.

2.6.1 Multiple comparisons techniques

The techniques outlined in this section are those that are the most relevant from the practical point of view. Of importance are the Tukey, Scheffe, and Bonferroni techniques. Miller[6] has an excellent presentation of the techniques discussed here.

Tukey method

The Tukey method of multiple comparisons refers to the test of all pairwise functions among effects. Thus, given the effects, one selects the pairwise function of interest, which we denote by L_T, and proceeds by finding the estimate of L_T.

Hence, with $L_T = \beta_i - \beta_j (i \neq j)$, the estimate is obtained by replacing the parameters by their estimators.

The confidence interval in this case is defined by $T s(\widehat{L}_T)$, where T is the upper 100α percent point of the studentized range with r and $n - r$ degrees of freedom and is denoted by $(1/\sqrt{2})q(1 - \alpha, r, n - r)$, and $s(\widehat{L}_T)$ is an estimate of the variance of $\widehat{L}_T$.

Note that in the preceding terminology, n is the total number of observations and r is the number of independent parameters in the model. The estimate $s(\widehat{L}_T)$ of the variance depends on the specific model. As an example, if the one-way layout with row effects is adopted, then the variance estimate is $SS_e (\mathbf{y}, \beta)/(n-r)$, where $n - r = m(n - 1)$.

Scheffe method

The Scheffe method of multiple comparisons refers to the test of all possible contrast functions given the effects. The confidence interval for a contrast function, denoted by L_S, in the case where we use q contrast functions, is $(qS)^{1/2}s(\widehat{L}_S)$, where S is the upper 100α percent point of the F-distribution with q and $n - r$ degrees of freedom and is denoted by $F_{\alpha,q,n-r}$, and $s(\widehat{L}_S)$ is an estimate of the variance of $\widehat{L}_S$. A detailed analysis of the Scheffe method follows in Section 2.6.3.

Bonferroni method

The Bonferroni method of multiple comparisons among effects applies to cases where we have pairwise, linear combinations, or contrast functions. For a class of q linear combinations, the confidence interval for a linear combination function, denoted by L_B, is $Bs(\widehat{L}_B)$, where B is the $100\,\alpha/2q$ percent point of the t-distribution with $n - r$ degrees of freedom and is denoted by $t(\alpha/2q, n - r)$ and $s(\widehat{L}_B)$ is the variance of $\widehat{L}_B$.

Although there are numerous other comparison techniques such as the Duncan test and least significant difference test (Fisher), only the Tukey and Scheffe methods have been used in image processing. This stems from the fact that the statistical tests used until recently, within the context of experimental designs, were primarily based on the F-test. The problem that frequently arises is the lack of knowledge about the specific location of the means for which the null hypothesis was rejected. The Scheffe method provides an answer to this question and, more importantly, is directly related to the F-test.[2]

In the next section, we present some motivation for the selection of an appropriate method for comparing two or multiple means.

2.6.2 Selection of a comparison method

Since all three methods seem to be appropriate for use in the image analysis problem, it is quite interesting to compare them. Apart from other considerations such as the sample size, model fitting, we concentrate on the only parameter that is of importance in our view, namely, the confidence interval for the estimator of the function of the effects. This is true because of the nature of the problem where the ultimate goal is generally the decision on the presence or absence of a feature given a certain confidence level.

- Pairwise comparisons

    If only pairwise comparisons are of interest, then the Tukey method gives better results given that it provides narrower confidence intervals, thus reducing false alarm errors.

- Contrast and linear combination comparisons

    Generally, the Bonferroni method gives better results than the Scheffe method. However, the Scheffe method has a rather interesting interpretation that if a test based on the F-statistic is rejected for a confidence level α, then a test based on the contrast function will provide means to locate the specific effect for which the test is rejected. Also, the Scheffe method is better than the Tukey method[4] in the sense that it gives narrower intervals.

Techniques based on the Scheffe method have already been developed by various researchers in image processing. In addition, the interpretation of the test in terms of

---

[4]When we use the modified Tukey method that handles the contrast function case.

F-statistics is of the utmost importance because it provides means for understanding the problem at hand from the practical point of view.

2.6.3 Estimable functions

In this section we review certain definitions of the concepts of estimable functions and contrasts with their associated confidence ellipsoids. An excellent presentation can be found in Scheffe.[2]

Definition 2.1 *A parametric function is defined as a linear function of the unknown parameters $\beta_1, \beta_2, \ldots, \beta_p$ where β_j $j = 1, \ldots, p$ is an effect defined under a particular design, with a known vector of constant coefficients $c_1, c_2, \ldots, c_p$.*

Denoting the parametric function by ψ, we have

$$\psi = \sum_{j=1}^{p} c_j \beta_j \tag{2.109}$$

In matrix notation, we can write Eq. (2.109) as

$$\psi = \mathbf{c}^T \beta \tag{2.110}$$

where $\mathbf{c}^T = (c_1, c_2, \ldots, c_p)$ and $\beta^T = (\beta_1, \beta_2, \ldots, \beta_p)$.

The following definition introduces the concept of estimable functions.

Definition 2.2 *A parametric function is called an estimable function if it has an unbiased linear estimate, that is, there exists a vector of constant coefficients such that*

$$\psi = E(\mathbf{a}^T \mathbf{y}) \tag{2.111}$$

no matter what the actual values of the effects are.

The following theorem provides means for the construction of the estimate of ψ and establishes the uniqueness of the estimate.

Theorem 2.1 (*Gauss–Markov*): *Under the assumption Ω of the general linear model, every estimable function $\psi = \mathbf{c}^T \beta$ has a unique unbiased linear estimate that has minimum variance in the class of all unbiased linear estimates. The estimate may be obtained from $\psi = \mathbf{c}^T \beta$ by replacing β by any set of least square estimates. If, in addition, the errors are normally* *distributed, then the estimate is of minimum variance among all unbiased estimators.*

Proof: See reference [2], p. 14.

2.6.4 Confidence ellipsoids and contrasts

Once the parametric function or contrast is formed, an interesting question arises how one should view the contrast from the statistical point of view since by virtue of the Gauss-Markov theorem, the contrast is an estimable function.

In the general case, we consider q linearly independent contrast functions that are generated from the p effects, that is, $\beta_j \; j = 1, \ldots, p$, using a matrix of coefficients $\mathbf{C}$, or

$$\psi = \mathbf{C}\beta \qquad (2.112)$$

In terms of the observations, the estimate is then

$$\widehat{\psi} = \mathbf{A}\mathbf{y} \qquad (2.113)$$

Lemma 2.1 $\widehat{\psi}$ *is* $N(\psi, \sum_{\widehat{\psi}})$ *where* $\sum_{\widehat{\psi}} = \sigma^2 \mathbf{B} = \sigma^2 \mathbf{A}\mathbf{A}^T$.

Proof: From the Gauss-Markov theorem, $E(\widehat{\psi}) = \psi$ and $E(\widehat{\psi}\widehat{\psi}^T) = AE(\mathbf{y}\mathbf{y}^T)\mathbf{A}^T = \sigma^2 \mathbf{A}\mathbf{A}^T$. Since the observations are independently normally distributed and from the linear relation (Eq. (2.113)), it follows that $\widehat{\psi}$ has a multivariate normal distribution with first and second order moments given by ψ and $\sigma^2 \mathbf{A}\mathbf{A}^T$, respectively.

Lemma 2.2 $\mathbf{B}$ *is nonsingular.*

Proof: $E(\widehat{\psi}) = \mathbf{A}E(\mathbf{y}) = \mathbf{A}\mathbf{X}^T\beta$, which is also equal to $\mathbf{C}\beta$ from Eq. (2.112). Hence, the matrix of weights $\mathbf{C}$ is identically equal to $\mathbf{A}\mathbf{X}^T$. But rank(C) is equal to q because we consider the contrasts to be independent. Hence, using the standard result of matrix algebra, $q = \text{rank}(\mathbf{C}) \le \text{rank}(\mathbf{A}) \le q$. The inequality is satisfied for rank($\mathbf{A}$) $= q$. Finally, rank($\mathbf{B}$) $= \text{rank}(\mathbf{A}\mathbf{A}^T) = \text{rank}(A)$, which implies that rank ($\mathbf{B}$) $= q$. Thus, $\mathbf{B}$ is nonsingular.

Theorem 2.2 *Under the* Ω *assumptions of the linear model,* $\widehat{\psi}$ *is* $N(\psi, \sum_{\widehat{\psi}})$ *and statistically independent of* $SS_e(\mathbf{y}, \beta)/\sigma^2$, *which is* χ^2_{n-r}.

Proof: That $SS_e(\mathbf{y}, \beta)/\sigma^2$ is χ^2_{n-r} follows from the fact that $E((\mathbf{y} - \mathbf{X}^T\beta)^T(\mathbf{y} - \mathbf{X}^T\beta)) = \sigma^2(n - r)$ and that an independent quadratic form, that is, $SS_e(\mathbf{y}, \beta)$, in independent standardized normal variates is a chi-square variate. Its degree of freedom is the rank of $SS_e(\mathbf{y}, \beta)$, which is $n - r$.[7]

To prove that $\widehat{\psi}$ and $SS_e(\mathbf{y}, \beta)/\sigma^2$ are independent, one can use an alternate description of the assumptions Ω of the linear model that is referred to as the canonical form [reference [2], p. 21].

Lemma 2.3 $(\widehat{\psi} - \psi)^T \mathbf{B}^{-1}(\widehat{\psi} - \psi)$ *is* $\sigma^2 \chi_q^2$ *and independent of* $SS_e(\mathbf{y}, \beta)$ $/\sigma^2$, *which is* χ_{n-r}^2.

Proof: See reference [2], Appendix V.

Theorem 2.3 *Under* Ω, *the probability is* $1 - \alpha$ *that*

$$(\widehat{\psi} - \psi)^T \mathbf{B}^{-1}(\widehat{\psi} - \psi) \leq q s^2 F_{\alpha, q, n-r} \tag{2.114}$$

where $s^2 = \widehat{\sigma}^2 = SS_e(\mathbf{y}, \beta)/(n - r)$.

Proof: The quotient $((\widehat{\psi} - \psi)^T \mathbf{B}^{-1}(\widehat{\psi} - \psi)/q)/(SS_e(\mathbf{y}, \beta)/(n - r))$ follows an F-distribution with q and $n - r$ degrees of freedom, which is denoted by $F_{q,n-r}$. Thus, given α, the probability is $1 - \alpha$ that $((\widehat{\psi} - \psi)^T \mathbf{B}^{-1}(\widehat{\psi} - \psi)/q)/(SS_e(\mathbf{y}, \beta)/(n - r)) \leq F_{\alpha, q, n-r}$. The important result shown above may be interpreted in terms of ellipsoids in the q-dimensional contrast space. The probability that the true vector ψ is contained in the ellipsoid is $1 - \alpha$ regardless of the parameter vector β.

2.7 Concluding remarks

In this chapter several important results from the theory of the ANOVA are presented. We first review the general problem of parameter estimation and linear hypothesis testing. We then describe two specific models of ANOVA; namely, the one-way and two-way designs. Their application in image processing is stressed by relating the effects to physical features such as column and row orientations.

Next, we present three commonly used incomplete designs; namely, the Latin squares, the Græco-Latin squares, and the SBIB designs. We then present some results from the theory of the multicomparison techniques.

Some important results associated with the Scheffe technique are elaborated upon. Whenever the proofs are omitted, the reader is referred to reference [2], which contains an excellent presentation of the technique.

3

Line detection

3.1 Introductory remarks

The extraction of line features from two-dimensional digital images has been a topic of considerable interest in the past two decades due to its numerous applications in astronomy, remote sensing, and medical fields. For example, typical problems in astronomy are the extraction of streaks corresponding to the trajectory of meteorites or satellites in space. In remote sensing, a major concern is to decipher from satellite images the network of roads and the separation among fields in agriculture. A common problem in both cases is the nature of the scene itself, which is often noisy with a complex background structure.

Among the techniques used in line extraction are those based on the matched filter concept. Under favorable noise conditions, namely, high SNR and i.i.d. Gaussian samples, the matched filter performs well. Unfortunately, typical noise conditions differ from the Gaussian distribution or, even if Gaussian, their variance is unknown. In addition, the SNR is usually not too high and varies unpredictably over the scene under consideration. As a result, a simple thresholding scheme will fail under these conditions.

Finally, the presence of structured backgrounds such as clouds or smoke will often hide parts of the line, and it is important to remove the background interference without affecting the discontinuity of the line.

In this chapter we present techniques for the detection of lines of various orientation based on the ANOVA models that were described in Chapter 2. As it is recalled, these techniques are UMP in Gaussian noise with unknown variance and display robustness if the background noise deviates from Gaussian. The procedures introduced here can be cast within the framework of hypothesis testing; namely, the model parameters representing some physical features are zero as compared to the alternative that they differ from zero. The test statistics are primarily based on the F-statistics within the framework of ANOVA. The specific orientations that are considered here are the horizontal, vertical, and diagonal directions or, equivalently,

$0°$, $45°$, $90°$, and $135°$ directions. We assume that the effects have already taken place, thus preventing any outside control. At this stage it is important to stress that, unlike in the regression analysis case, ANOVA models assume that the effect is either present or absent. Therefore, the design matrix $\mathbf{X}^T$ in the linear model contains only 1s and 0s. The chapter is organized as follows.

We first explore the one-way design in the unidirectional line detection problem. It is shown that the procedure can only be used for the discrimination in one direction at a time. Next, we extend the technique to the two-way design, highlighting the advantage in using this particular design over the unidirectional design, specifically in the number of simultaneous directions that can be detected at any one time. We further extend the procedure to include the four possible orientations by using a Græco-Latin design. Next, we consider the multidirectional detector based on the contrast function methodology. An extension of the theory to include the correlated noise case under the assumption of a first-order Markov noise process is then presented. The important problem of trajectory detection is accomplished through the use of incomplete designs. Finally, adaptive detectors based on the GLS design, useful for line detection in structured backgrounds, are presented.

3.2 Unidirectional line detectors

In this section we assume that only one effect, either horizontal or vertical, has already taken place. In line detection terminology it corresponds to the presence of either horizontally or vertically oriented line segments within a localized region of the scene. The data is collected from the image using a mask of dimensions $m \times n$. We assume that $m \ll I$ and $n \ll J$, where I and J are the vertical and horizontal dimensions of the image. Typical values for m and n range from 3×3 to 7×7 pixels. The observations are parameterized using the model in Eq. (2.18), where for the sake of clarity row effects are chosen to represent the parameters of interest. In addition, we assume for the time being that the noise samples are Gaussian i.i.d with zero mean and variance σ^2. Rewriting Eq. (2.18) and including the side condition, we have

$$\Omega : \begin{cases} y_{ij} = \mu + \alpha_i + e_{ij} & i = 1, 2, \ldots, m; j = 1, 2, \ldots, n \\ \sum_{i=1}^{m} \alpha_i = 0 \\ (e_{ij}) \text{ independent } N(0, \sigma^2 \mathbf{I}) \end{cases} \quad (3.1)$$

Consider that a window over a certain region of the image has a structure similar to that shown in Fig. 3.1. The problem is thus reduced to the test of hypothesis H_a : all $\alpha_i = 0$, where each α_i, $i = 1, \ldots, m$ represents a horizontal line.

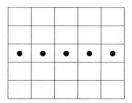

Figure 3.1. 5×5 window with horizontal line.

Using Eq. (2.28) with the threshold corresponding to the one-way design, the statistic for testing H_a for a level of the test α is

$$F_a = \frac{\sum_{i=1}^{m} \sum_{j=1}^{n} (y_{i.} - y_{..})^2 / (m-1)}{\sum_{i=1}^{m} \sum_{j=1}^{n} (y_{ij} - y_{i.})^2 / m(n-1)} \tag{3.2}$$

Since the column index j does not appear in the numerator of Eq. (3.2), a further reduction of Eq. (3.2) is possible, yielding

$$F_a = \frac{n \sum_{i=1}^{m} (y_{i.} - y_{..})^2 / (m-1)}{\sum_{i=1}^{m} \sum_{j=1}^{n} (y_{ij} - y_{i.})^2 / m(n-1)} \tag{3.3}$$

If F_a exceeds the threshold $F_{\alpha, m-1, m(n-1)}$, the hypothesis is rejected. Although simple, the test proposed above is powerful enough to enable the detection of lines oriented in the horizontal direction within the boundaries of the scanning window.

The implementation of the detector is summarized in the following steps.
- Calculate the mean estimate $\hat{\mu}$.
- Calculate effect estimates $\hat{\alpha}_i$, $i = 1, 2, \ldots, m$.
- Find the "between sum of squares" (BSS).
- Find the "within sum of squares" (WSS).
- Form the test statistic by taking the ratio of the weighed sum of squares.

Of course, as we mentioned in the beginning, effects that tend to align in the vertical direction are rejected using Eq. (3.3). An extension of the detection procedure to the detection of vertical lines will proceed with a parameterization based on column effects. Thus, the hypothesis becomes H_a : all $\beta_j = 0$, which translates in the test of whether vertical lines are present or not. Consequently, the test statistic is

$$F_b = \frac{m \sum_{j=1}^{n} (y_{.j} - y_{..})^2 / (n-1)}{\sum_{i=1}^{m} \sum_{j=1}^{n} (y_{ij} - y_{.j})^2 / n(m-1)} \tag{3.4}$$

The scanning for the test of both vertical and horizontal lines can be done concurrently using two masks working in parallel. Fig. 3.3 shows the result obtained by processing the image in Fig. 3.2. As is shown, in addition to being detected, the line is also thickened. To understand the origin of this problem, consider the set of means θ such that $\theta = \{\alpha_1, \alpha_2, \ldots, \alpha_m\}$. The hypothesis H_a : all $\alpha_i = 0$ is now

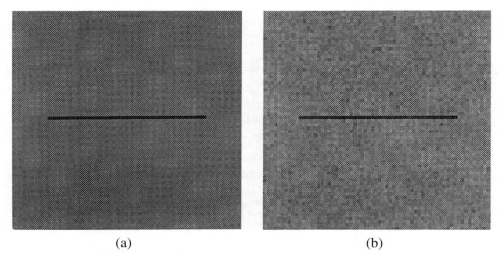

(a)   (b)

Figure 3.2. (a) Horizontal line in uniform background, (b) noisy image,
$\sigma = 8$ (SNR $\simeq$ 10 dB).

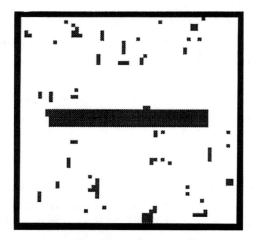

Figure 3.3. Output of one-way detector, $\alpha = 5\%$.

rephrased in the following alternate form

$$H : \theta = 0$$
$$K : \theta \neq 0$$

(3.5)

Eq. (3.5) is the classic hypothesis-alternative pair in which H corresponds to the means of the rows in the window being equal to zero and therefore represents the cases in which no structure is present. By structure, we mean the line target of interest or any other arbitrary background structure oriented in the direction of the rows. Therefore, if the pattern to be detected is a horizontal line one-pixel wide through the center of the window, it would appear as a shift in the mean of the central row. Thus, in terms of the hypothesis-alternative pair, we can write

$$H : \theta = 0$$
$$K : \alpha_1 = \alpha_2 = \alpha_4 = \alpha_5 \neq \alpha_3 \tag{3.6}$$

Carrying this idea further, we see that a line not centrally located within the window can be modeled as a shift in the mean of the appropriate row location. Consequently, the space of the alternative of interest as expressed by K is only a subspace of the alternative subspace $\Omega - \omega$ in Eq. (2.16). As a result, the statistic in Eq. (3.3) tests for more alternatives than is necessary, which leads to the conclusion that the conditions of the UMP-invariant test are not met.

A solution to this problem is to use the multicomparison test based on Scheffe test. Habestroh[16] proposed a shape test in terms of the effects estimates. To see that clearly, assume that $m = n = 5$ and the pattern of interest is a line running across the center of the window. The shape statistic is then

$$s(\alpha) = 4\widehat{\alpha}_3 - (\widehat{\alpha}_1 + \widehat{\alpha}_2 + \widehat{\alpha}_4 + \widehat{\alpha}_5) \tag{3.7}$$

Note that the shape test has the general form of a contrast function because the coefficients sum up to zero. In addition, the shape statistic is tested against a threshold set to zero. A more conventional threshold in the Scheffe comparison case is equal to

$$(\sigma^2 \mathbf{a}\mathbf{a}^T F_{\alpha,1,n-r})^{1/2} \tag{3.8}$$

It can be argued that using a threshold in the comparison increases the processing time. However, with the advent of high-speed processors, the difference in computation time is without serious consequence on the overall performance of the detector. Fig. 3.4 demonstrates the improvement obtained by including the shape test in the detection process.

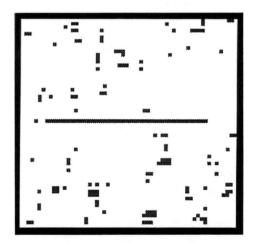

Figure 3.4. Output of one-way detector coupled with a shape test.

3.3 Bidirectional line detectors

Detectors based on the one-way ANOVA model are sensitive only to the orientation for which they were designed. This in itself constitutes a very serious drawback to their use in the general case because it is unrealistic to assume that real images contain only one-line orientation. Although a bank of two unidirectional detectors, each designed for a specific orientation, will solve this problem at the expense of overall simplicity and processing time, a relatively simple approach in the form of a two-way design is preferred.

In this section we introduce bidirectional detectors based on the two-way design. We first derive the detector structure. We then highlight the advantages in using a bidirectional detector over the unidirectional detector.

The model chosen to represent the orthogonal directions is given by Eq. (2.29),

$$\Omega : \begin{cases} y_{ij} = \mu + \alpha_i + \beta_j + e_{ij} & i = 1, 2, \ldots, m; j = 1, 2, \ldots, n \\ \sum_{i=1}^{m} \alpha_i = 0; \ \sum_{j=1}^{n} \beta_j = 0. \\ (e_{ij}) \text{ independent } N(0, \sigma^2 \mathbf{I}) \end{cases} \tag{3.9}$$

Horizontal lines are represented by the row effects α_i, $i = 1, 2, \ldots, m$, whereas vertical lines are represented by the column effects β_j, $j = 1, 2, \ldots, n$. Comparing Eqs. (3.9) to (3.1), we notice that the additional parameters β_j are used to formulate the hypothesis-alternative pair on the column effects. Thus, the two-way ANOVA model provides a framework for simultaneous detection of horizontal and vertical lines. In terms of the α_i and β_j, the hypotheses are

$$\begin{aligned} H_a &: \text{ all } \alpha_i = 0 \\ H_b &: \text{ all } \beta_j = 0 \end{aligned} \tag{3.10}$$

and the test of the hypotheses is carried out using

$$F_a = \frac{n \sum_{i=1}^{m} (y_{i.} - y_{..})^2 / (m-1)}{\sum_{i=1}^{m} \sum_{j=1}^{n} (y_{ij} - y_{i.} - y_{.j} + y_{..})^2 / (n-1)(m-1)} \tag{3.11}$$

and

$$F_b = \frac{m \sum_{j=1}^{n} (y_{.j} - y_{..})^2 / (n-1)}{\sum_{i=1}^{m} \sum_{j=1}^{n} (y_{ij} - y_{i.} - y_{.j} + y_{..})^2 / (n-1)(m-1)} \tag{3.12}$$

Using a square window ($m = n$), the detector structure is obtained by considering the maximum value of the pair (F_a, F_b) and comparing it to the threshold T_H, where $T_H = F_{\alpha,(m-1),(m-1)(n-1)}$. The detector structure is then

$$f_d \mathop{\gtrless}^{<}_{>} T_H \tag{3.13}$$

where $f_d = \max(F_a, F_b)$. The implementation of the detector can be summarized using the following steps.

- Calculate the mean estimate $\widehat{\mu}$.
- Calculate effect estimates $\widehat{\alpha}$, $i = 1, 2, \ldots, m$ and $\widehat{\beta}$, $j = 1, 2, \ldots, n$.
- Find the BSS_1 and BSS_2.
- Find the *WSS*.
- Form the test statistic by taking the ratio of the weighed sum of squares.
- Determine $f_d = \max(F_a, F_b)$.

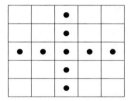

Figure 3.5. 5×5 window for bidirectional detection of line segments.

Fig. 3.6 shows the output image using Eq. (3.13). As explained before, the thickening effect is due primarily to the test of more alternatives than is necessary. Coupling Eq. (3.13) with a shape test will eliminate this problem. With $U(.)$ denoting the step function, we use the following structure

$$f_d = \max_{k=a,b} F_k U(s_k) \tag{3.14}$$

where $s_k = 4\gamma_3 - (\gamma_1 + \gamma_2 + \gamma_4 + \gamma_5)$ and γ denotes either row or column effects depending on the shape under consideration. The detector output is then based on

$$f_d \underset{>}{\overset{<}{}} T'_H \tag{3.15}$$

where $T'_H = F_{2\alpha,(m-1),(m-1)(n-1)}$.

Fig. 3.7 is obtained using Eq. (3.15). Although a principal advantage of the two-way ANOVA based detector over the one-way based detector is the simultaneous detection in both the $0°$ and $90°$ directions, a subtle and less apparent advantage is in the possibility of modeling background structures with the present design. To see this, assume that we have a horizontal line crossing a vertical discontinuity as shown in Fig. 3.8. First, suppose that the one-way based detector with row effect as parameters is used to detect the line. In the vicinity of the edge, the difference in column mean estimates contributes to the denominator component of Eq. (3.3). But because $SS_e(\mathbf{y}, \boldsymbol{\beta})/(n-r)$ represents the noise variance estimate within the window,

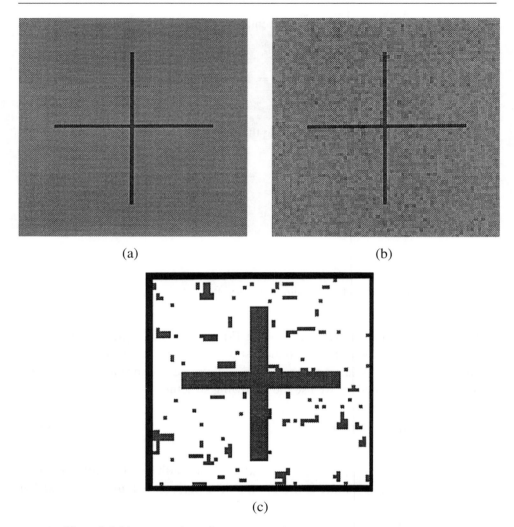

(a)   (b)

(c)

Figure 3.6. Line processing using a two-way detector. (a) Clean image, (b) noisy image, $\sigma = 8$ (SNR $\simeq$ 10 dB), (c) output of two-way detector, $\alpha = 5\%$.

it implies that the variance estimate increases due to the background structure rather than due to the noise effect.

Accordingly, the power of the test is reduced due to the decrease of the F-statistic value. Fig. 3.9 shows the result of applying the procedure to the image in Fig. 3.8. Note the presence of a discontinuity in the vicinity of the edge. On the other hand, if a two-way based detector is used on the same image, the difference in intensity levels contributes only to the "column sum of squaress" BSS_2, which does not appear in the F-statistic corresponding to row effects. Hence, it is possible to

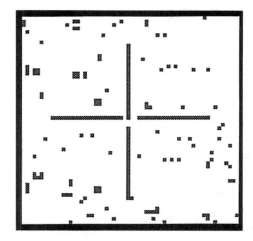

Figure 3.7. Output of two-way detector coupled with shape tests for horizontal and vertical lines.

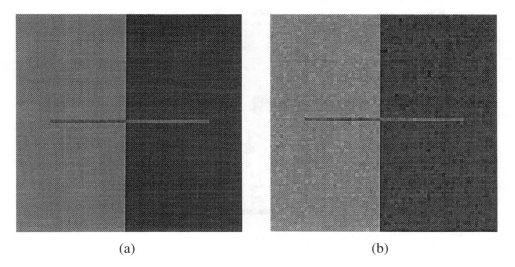

(a)   (b)

Figure 3.8. Line crossing a vertical edge. (a) Clean image, (b) noisy image, $\sigma = 8$.

suppress this type of structure completely using the two-way approach. Note that the degrees of freedom allocated to WSS are reduced by the number allocated to BSS_2 (due to the column effects).

Fig. 3.10 shows the result of applying the two-way based ANOVA detector to the image in Fig. 3.8. In a similar manner, a horizontally oriented intensity discontinuity will be rejected using the F-statistic corresponding to column effects.

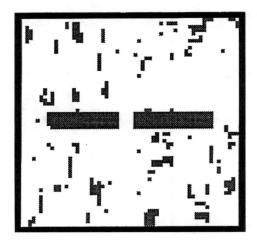

Figure 3.9. Output of one-way detector, $\alpha = 5\%$.

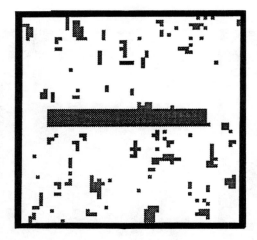

Figure 3.10. Output of two-way detector, $\alpha = 5\%$.

3.4 Multidirectional line detectors

Although the detection of lines oriented in the $0°$ and $90°$ directions is sufficient in most cases, it is interesting to extend the bidirectional detector to include the diagonal directions, that is, the $45°$ and $135°$ diagonally oriented lines. The design of interest in this case is the Græco-Latin square layout introduced in Section 2.5.2. A typical 5×5 layout is shown in Fig. 2.4. Note that the design is the nonrandomized version of the Græco-Latin square, that is, the arrangement is such that every treatment is coupled exactly once with every treatment. The treatments in Fig. 2.4

are the rows, columns, and Greek and Latin letters. The model in this case allows the simultaneous detection in any of the four distinct directions.

Recalling Eq. (2.68) with the usual side conditions, we have

$$\Omega : \begin{cases} y_{ij} = \mu + \alpha_i + \beta_j + \tau_k + \delta_l + e_{ij} \quad (i, j, k, l) \in S \\ \sum_{i=1}^{m} \alpha_i = 0; \sum_{j=1}^{m} \beta_j = 0; \sum_{k=1}^{m} \tau_k = 0; \sum_{l=1}^{m} \delta_l = 0. \\ (e_{ij}) \ is \ N(0, \sigma^2 \mathbf{I}) \end{cases} \quad (3.16)$$

The four hypotheses-alternative pairs to be tested are in this case

$$\begin{aligned} H_a &: \alpha = 0 & K_1 &: \alpha \neq 0 \\ H_b &: \beta = 0 & K_2 &: \beta \neq 0 \\ H_c &: \tau = 0 & K_3 &: \tau \neq 0 \\ H_d &: \delta = 0 & K_4 &: \delta \neq 0 \end{aligned} \quad (3.17)$$

where α, β, τ, and δ are such that $\alpha = \{\alpha_1, \alpha_2, \ldots, \alpha_m\}$, $\beta = \{\beta_1, \beta_2, \ldots, \beta_m\}$, $\tau = \{\tau_1, \tau_2, \ldots, \tau_m\}$, and $\delta = \{\delta_1, \delta_2, \ldots, \delta_m\}$. The F-statistics for testing H_a, H_b, H_c, and H_d are given in Eqs. (2.76), (2.77), (2.78), and (2.78). A major advantage of the present model is that the detection of a line of a given orientation can be done independently of the existence or nonexistence of a line in any other direction. In this case, we say that the F-statistics are independent of each other or mutually decoupled.

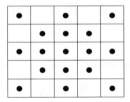

Figure 3.11. 5×5 window for multidirectional detection of line elements.

One possible structure for the detector is comprised of the maximum F-statistic among those for which the central treatment effect estimate is greater than the overall average. In this case, we have

$$f_d = \max_{m=1,2,3,4} F_m U(s_m) \quad (3.18)$$

where $s_m = 4\gamma_3 - (\gamma_1 + \gamma_2 + \gamma_4 + \gamma_5)$ and γ denotes either row, column, or diagonal effects depending on the direction under consideration. The output is based on

$$f_d \underset{>}{\overset{<}{}} T_H \quad (3.19)$$

where $T_H = F_{\alpha', (m-1), (m-1)(m-3)}$ and $\alpha' = 2[1 - (1 - \alpha)^{1/4}]$. The implementation of the detector can be summarized using the following steps.

- Calculate the mean estimate $\widehat{\mu}$.
- Calculate effect estimates $\widehat{\alpha}_i$, $\widehat{\beta}_j$, $\widehat{\tau}_k$, $\widehat{\delta}_l$.
- Find the BSS_1, BSS_2, BSS_3, and BSS_4.
- Find the WSS.
- Form the test statistics by taking the ratio of the corresponding weighed sum of squares.
- Find the shape statistics.
- Find max $F_m U(s_m)$, $m = 1, 2, 3, 4$.

Figs. 3.12 and 3.13 show the output images using the basic GLS detector and the detector based on Eqs. (3.18) and (3.19).

Additional benefits of using the GLS detector as compared to the low-order ANOVA based detector are in the modeling of more nonuniform background structures in directions that are statistically independent of the occurrence of the line of interest. This was shown in references [16] and [17] to alleviate some of the problems associated with the detection of lines embedded in structured backgrounds.

3.5 Multidirectional contrast detectors

Until now, the detector structure we assumed was rather simple in that the line detection was based on a simple F-test in conjunction with a shape test. Notwithstanding the simplicity of the structure, the overall performance of the line detectors is unusually good. Indeed, the detectors presented in the preceding sections achieve good results even in low SNR in addition to poor contrast in the case of structured

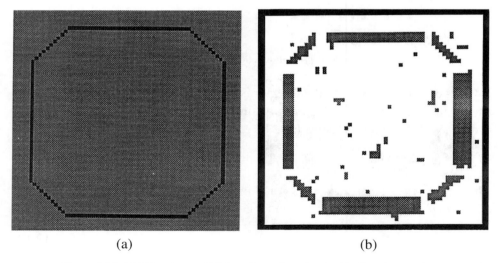

(a)   (b)

Figure 3.12. Multiline image (All four line orientations). (a) Clean image, (b) output of basic GLS detector, $\alpha = 5\%$.

background. The shape test solves the line thickening problem.

However, the threshold used, in this case zero, may be in some cases irrelevant or insufficient for a given level of the test α. It is then necessary to use a threshold without undue increase in complexity of the detector.

The most common structure in this case is based on the contrast function methodology. In terms of the observations within the window, what we are interested in finding is whether there is a significant visual contrast between the line and the surrounding background. In statistical terms the process translates to finding an estimate of the contrast function and the associated confidence interval. The procedure would enable us to determine in a hypothesis-alternative setting whether the contrast function assigned to the window in question is zero with a certain confidence level. In the unidirectional and bidirectional line detector cases, the contrasts of interest in terms of the row and column effects are

- Row contrast function

$$\widehat{\psi}_{0^\circ} = 4\widehat{\alpha}_3 - (\widehat{\alpha}_1 + \widehat{\alpha}_2 + \widehat{\alpha}_4 + \widehat{\alpha}_5) \tag{3.20}$$

- Column contrast function

$$\widehat{\psi}_{90^\circ} = 4\widehat{\beta}_3 - (\widehat{\beta}_1 + \widehat{\beta}_2 + \widehat{\beta}_4 + \widehat{\beta}_5) \tag{3.21}$$

Note that these contrasts are efficient at locating lines running through the center of the mask. In addition, it is possible to use additional contrast functions that are orthogonal to Eqs. (3.20) and (3.21), respectively. We use what is referred to as orthogonal contrasts with their total number being fixed by the maximum number of effects. For example, there are at most four orthogonal contrasts for five effects

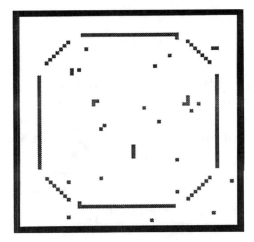

Figure 3.13. Output of basic GLS detector coupled with shape tests for all four orientations.

in a window. In the present case it is sufficient to use only one contrast function for each parameter. Hence, we select $q = 1$, which reduces to the single comparison case of Section 2.6.4. The most natural way of representing the effects in terms of physical features is to assume the average of all pixel values along a row as representing the desired row effect. A similar reasoning also applies to column effects. Thus, the contrasts in Eqs. (3.20) and (3.21) become

- Center row contrast

$$\widehat{\psi}_{0°} = \frac{4}{5} y_{3.} - \frac{1}{5}(y_{1.} + y_{2.} + y_{4.} + y_{5.}) \tag{3.22}$$

- Center column contrast

$$\widehat{\psi}_{90°} = \frac{4}{5} y_{.3} - \frac{1}{5}(y_{.1} + y_{.2} + y_{.4} + y_{.5}) \tag{3.23}$$

With the present line characterization, we are led to the conclusion that the least squares estimates of the effects are identical to the sample mean effect estimates. Turning to the multidirectional case, one should remark that it is impossible to relate the effects to the physical line definition based on the specific GLS structure shown in Fig. 2.4.

The estimates of treatments represented by the Greek and Latin letters in Eq. (2.68) are not readily associated with the diagonal lines. Therefore, to be consistent with the contrast definition for the row and column cases, a window with a parallepiped shape is used. This allows the inclusion of additional pixels in the off-center lines, which permits the use of the sample mean estimates. Consequently, the diagonal masks are defined as follows

- Center 45° diagonal

$$\widehat{\psi}_{45°} = \frac{4}{5}\sum_{j=1}^{n} y_{j,(m-j+1)} - \frac{1}{5}\left[\sum_{j=1}^{n} y_{j,(m-j-1)} + \sum_{j=1}^{n} y_{j,(m-j+2)}\right.$$
$$\left. + \sum_{j=1}^{n} y_{j,(m-j+3)} + \sum_{j=1}^{n} y_{j,(m-j)}\right] \tag{3.24}$$

- Center 135° diagonal

$$\widehat{\psi}_{135°} = \frac{4}{5}\sum_{j=1}^{n} y_{jj} - \frac{1}{5}\left[\sum_{j=1}^{n} y_{j,(j-1)} + \sum_{j=1}^{n} y_{j,(j-2)}\right.$$
$$\left. + \sum_{j=1}^{n} y_{j,(j+1)} + \sum_{j=1}^{n} y_{j,(j+2)}\right] \tag{3.25}$$

Although there are four distinct windows, in practice we use only three masks. The first rectangular window is used to determine the center row and column contrasts. The remaining parallepiped masks are used for the calculation of the 45° and 135° diagonal contrasts.

For the multicomparison model, an essential parameter is the noise variance estimate, which is equal to $SS_e(\mathbf{y}, \beta)/(n-r)$. Therefore, it is necessary to determine the sum of squares under the alternative. Note that the degree of freedom $n - r$ is equal to $(m - 1)(m - 3)$ for the GLS design. Recalling Eq. (2.72), we have

$$s^2 = \sigma^2 = \frac{\sum_{(i,j,k,l)\in S}(y_{ij} - y_{i...} - y_{.j..} - y_{..k.} - y_{...l})^2}{(m - 1)(m - 3)} \tag{3.26}$$

Because we consider the one-contrast case ($q = 1$) for the four directional contrast, Eq. (2.116) reduces to

$$(\widehat{\psi} - \psi)^2 \leq s^2 \mathbf{a}\mathbf{a}^T F_{\alpha,1,(m-1)(m-3)} \tag{3.27}$$

where $\mathbf{a}$ is the vector of coefficients from Eq. (2.115). The contrast variance is then

$$\widehat{\sigma}_{\widehat{\psi}}^2 = s^2 \mathbf{a}\mathbf{a}^T \tag{3.28}$$

To determine the components of $\mathbf{b}$ in Lemma 2.1, consider, for example, Eq. (3.23). The elements of $\mathbf{a}$ are the coefficients assigned to each effect in the window. For example, for $m = 5$,

$$a_{ij} = \begin{cases} 4/5 & i = 1, 2, \ldots, 5; j = 5(i-1) + 3. \\ -1/5 & \text{otherwise} \end{cases} \tag{3.29}$$

With the present structure, we normally do not need to find the F-statistics corresponding to the four hypotheses-alternative pairs. Instead, the test is carried out by considering C_d such that

$$C_d = \max(\widehat{\psi}_{0°}, \widehat{\psi}_{90°}, \widehat{\psi}_{45°}, \widehat{\psi}_{135°}) \tag{3.30}$$

Then, the corresponding test is of the form

$$|C_d| \underset{>}{\overset{<}{}} (S\widehat{\sigma}_{\widehat{\psi}_i}^2)^{1/2} \tag{3.31}$$

where

$$S = F_{\alpha,1,(m-1)(m-3)} \tag{3.32}$$

A line is declared present whenever C_d exceeds the threshold on the right-hand side of Eq. (3.31). The maximum contrast corresponds to the line orientation with the strongest contrast.

The implementation of the detector can be summarized using the following steps.

- Calculate all four contrasts.
- Determine the maximum contrast.
- Find the corresponding contrast variance.
- Determine the threshold.
- Compare the maximum contrast to the threshold.

Fig. 3.14 shows the results obtained by using the contrast detector on the image shown in Fig. 3.12(a).

3.6 Multidirectional detectors in correlated noise

Until now, we discussed the detection of lines in scenes where the noise samples are Gaussian i.i.d. In the present section we treat the correlated noise case assuming

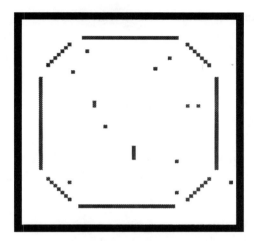

Figure 3.14. Output of basic GLS detector coupled with shape tests for all four orientations.

Markov dependence of the first order. The contrast-based detector is retained for the extraction and location of the possible lines. Although the F-test derived from ANOVA is robust with regard to the deviation from the underlying Gaussian distribution, robustness does not readily extend to the case where there is correlation among image pixels. This is due to the appearance of localized patterns that are solely due to the additive corrupting noise. As a result, these false patterns may interfere with the detection process by either masking the line of interest or yielding characteristics similar to a line. The former problem seems to occur more frequently than the latter and only in high noise correlated environments.

A viable mathematical model that enables the introduction of line detection in correlated noise is based on the Markov process. A commonly used model for stationary correlated noise in which the correlation factor is defined as

$$r(i_2 - i_1, j_2 - j_1) = E(e_{i_1 j_1} e_{i_2 j_2})/\sigma^2 \tag{3.33}$$

where (e_{ij}) is the zero mean additive noise for the pixel position (i, j) and σ^2 is the noise variance, is given by

$$r(N) = r_o^N \quad N = |i_2 - i_1| + |j_2 - j_1| \tag{3.34}$$

N is the distance measured in pixels between pixel locations (i_1, j_1) and (i_2, j_2). The dependency among pixels is reflected in the introduction of the correlation matrix $\mathbf{K}_f$ such that $E(\mathbf{e}\mathbf{e}^T) = \sigma^2 \mathbf{K}_f$.

The observations within the scanning window can be represented by a matrix $\mathbf{Y}$ with dimensions $m \times n$. In matrix notation, we have

$$
\mathbf{Y}_{m \times n} = \begin{pmatrix} y_{11} & y_{12} & \cdots & y_{1n} \\ y_{21} & y_{22} & \cdots & y_{2n} \\ \vdots & \vdots & \cdots & \vdots \\ y_{m1} & y_{m2} & \cdots & y_{mn} \end{pmatrix}
\tag{3.35}
$$

Using a row or column scanning, or a linear stacking operator,[18] Eq. (3.35) is converted to a vector $\mathbf{y}$, which facilitates the derivation of $\mathbf{K}_f$. Hence, we can write

$$
\mathbf{K}_f = E(\mathbf{y}\mathbf{y}^T)
\tag{3.36}
$$

A reasonable and accurate model for typical image scenes described in reference [18], is the tensored correlated model (TCM), that is, the correlation between elements is separable into the product of row and column correlation functions. From Eq. (3.34), the correlation coefficient is

$$
r(i_2 - i_1, j_2 - j_1) = \rho_r^{|i_2 - i_1|} \rho_c^{|j_2 - j_1|}
\tag{3.37}
$$

and Eq. (3.36) reduces to $\mathbf{K}_f = \mathbf{K}_r \otimes \mathbf{K}_c$, where

$$
\sigma^2 \mathbf{K}_r = \sigma_r^2 \begin{pmatrix} 1 & \rho_r & \cdots & \rho_r^{n-1} \\ \rho_r & 1 & \cdots & \rho_r^{n-2} \\ \vdots & \vdots & \cdots & \vdots \\ \rho_r^{n-1} & \rho_r^{n-2} & \cdots & 1 \end{pmatrix}
\tag{3.38}
$$

and

$$
\sigma^2 \mathbf{K}_c = \sigma_c^2 \begin{pmatrix} 1 & \rho_c & \cdots & \rho_c^{m-1} \\ \rho_c & 1 & \cdots & \rho_c^{m-2} \\ \vdots & \vdots & \cdots & \vdots \\ \rho_c^{m-1} & \rho_c^{m-2} & \cdots & 1 \end{pmatrix}
\tag{3.39}
$$

with ρ_r being the correlation coefficient between two adjacent row pixels. Similarly, ρ_c is the correlation coefficient between two adjacent column pixels. Note that the inverse of a TCM matrix is also a TCM matrix, that is, $\mathbf{K}_f^{-1} = \mathbf{K}_r^{-1} \otimes \mathbf{K}_c^{-1}$.

An essential component of the contrast procedure of Section 3.5 is the contrast variance, which for the independent noise case is given by Eq. (3.28). For the dependent case we have to consider the matrix $\mathbf{K}_f$ in the derivation of $\widehat{\sigma}_{\psi}^2$. It has been shown in Eq. (2.115) that $\widehat{\psi} = \mathbf{A}\mathbf{y}$. For the one-comparison case ($q = 1$) the contrast variance is

$$\text{Var}(\widehat{\psi}) = E\left(\sum_i \sum_j a_{ij} y_{ij} \sum_i \sum_j a_{ij} y_{ij}\right)$$
$$= \sigma^2 \sum_i \sum_j \sum_k \sum_l a_{ij} a_{kl} \rho_{ijkl} \tag{3.40}$$

where ρ_{ijkl} is the correlation coefficient between two pixels at locations (i, j) and (k, l) in the matrix representation. Eq. (3.40) can also be written in vector form as

$$\text{Var}(\widehat{\psi}) = \sigma^2 \mathbf{a}^T \mathbf{K}_f \mathbf{a} \tag{3.41}$$

where σ^2 is estimated by $SS_e(\mathbf{y}, \beta)/(n - r)$. The error sum of squares $SS_e(\mathbf{y}, \beta)$ for the data decorrelated by $\mathbf{K}_f$ is given in Eq. (2.14); that is,

$$SS_e(\mathbf{y}, \beta) = (\mathbf{y} - \mathbf{X}^T \beta)^T \mathbf{K}_f^{-1} (\mathbf{y} - \mathbf{X}^T \beta) \tag{3.42}$$

The calculation of $\widehat{\sigma}_{\widehat{\psi}}^2$ as defined in Eq. (3.41) requires the knowledge of $\mathbf{K}_f$. Although the tensored form is valid for rectangular masks, it is not adequate for other masks. Recall that for the contrast procedure, we use three combined masks: a rectangular mask for the detection of horizontal and vertical lines and two diagonal masks for the detection of 45°- and 135°-oriented lines. In the latter case the correlation matrix form depends on the arrangement of the elements in the diagonal mask. We consider first the case of the rectangular mask for the TCM form. From Eq. (3.29) we have for a 5×5 window and a vertical contrast

$$a_{ij} = \begin{cases} 4/5 & i = 1, 2, \ldots, 5; \, j = 5(i - 1) + 3. \\ -1/5 & \text{otherwise} \end{cases} \tag{3.43}$$

Using Eq. (3.41), it can be shown that

$$\widehat{\sigma}_{\widehat{\psi}}^2 = \frac{\sigma^2}{25}(5 + 8\rho_r + 6\rho_r^2 + 4\rho_r^3 + 2\rho_r^4)$$
$$\cdot (20 - 12\rho_c - 14\rho_c^2 + 4\rho_c^3 + 2\rho_c^4) \tag{3.44}$$

Considering the diagonal masks, we first notice that the vector of coefficients $\mathbf{a}$ is identical to Eq. (3.43), which is clearly seen from the diagonal arrangement. The elements of the correlation matrix are in the form

$$K_f(i, j) = \rho_r^{|x_i - x_j|} \rho_c^{|y_i - y_j|} \tag{3.45}$$

The coordinates x_i and y_j are obtained directly from the pixel position in the diagonal masks with x and y denoting the horizontal and vertical positions, respectively.

For the 45°-oriented lines the vector positions are

$$\mathbf{x}^T = [(2\ 3\ 4\ 5\ 6)(1\ 2\ 3\ 4\ 5)(0\ 1\ 2\ 3\ 4)(-1\ 0\ 1\ 2\ 3)(-2\ -1\ 0\ 1\ 2)] \tag{3.46}$$

and

$$\mathbf{y}^T = [(0\ 0\ 0\ 0\ 0)(1\ 1\ 1\ 1\ 1)(2\ 2\ 2\ 2\ 2)(3\ 3\ 3\ 3\ 3)(4\ 4\ 4\ 4\ 4)] \tag{3.47}$$

Using Eq. (3.41) it can be shown that the contrast variance corresponding to the 45° and 135° diagonals is

$$
\begin{aligned}
\hat{\sigma}_{\hat{\psi}}^2 = \frac{\sigma^2}{25} \big[& 5(20 - 12\rho_c - 14\rho_c^2 + \rho_c^3 + 2\rho_c^4) \\
& + 8\rho_r(-6 + 13\rho_c - 4\rho_c^2 - 6\rho_c^3 + 2\rho_c^4 + \rho_c^5) \\
& + 6\rho_r^2(-7 - 4\rho_c + 21\rho_c^2 - 6\rho_c^3 - 7\rho_c^4 + 2\rho_c^5 + \rho_c^6) \\
& + 4\rho_r^3(2 - 6\rho_c - 6\rho_c^2 + 20\rho_c^3 - 6\rho_c^4 - 7\rho_c^5 + 2\rho_c^6 + \rho_c^7) \\
& + 2\rho_r^4(1 + 2\rho_c - 7\rho_c^2 - 6\rho_c^3 + 20\rho_c^4 - 6\rho_c^5 - 7\rho_c^6 + 2\rho_c^7 + \rho_c^8) \big]
\end{aligned}
\tag{3.48}
$$

Note that the correlation matrix corresponding to the 135° orientation is identical to the 45° orientation because of symmetry. It remains to calculate $SS_e(\mathbf{y}, \beta)$. To find β, the vector of parameters to be estimated, an alternative to the direct minimization of Eq. (3.42) has been suggested by Behar [14] and is presented next. To that end, consider the parametric model for the GLS design.

$$\mathbf{y} = \mathbf{X}^T \beta + \mathbf{e} \tag{3.49}$$

where $\mathbf{X}^T$ is the design matrix. Assume that rank$(\mathbf{X}^T) = r \leq p$, where p is the number of parameters in the design. Since $\mathbf{X}^T$ is usually not of full rank$(r < p)$, it is customary to add t side conditions on the p parameters β in the form $\mathbf{H}^T \tilde{\beta} = 0$. The elements of $\mathbf{H}$, where $\mathbf{H}^T$ is of dimensions $t \times p$ $(t \geq p - r)$, are known fixed constants. By imposing additional side conditions, we make the set of parameters β in Eq. (3.49) unique in the sense that there exists a unique set of parameters $\tilde{\beta}$ that satisfies $\mathbf{X}^T \beta = \mathbf{X}^T \tilde{\beta}$ and $\mathbf{H}^T \tilde{\beta} = 0$.

Now, consider the augmented matrix $\mathbf{G}^T$ such that

$$\mathbf{G}^T = \begin{pmatrix} \mathbf{X}^T \\ \mathbf{H}^T \end{pmatrix} \tag{3.50}$$

where rank$(\mathbf{G}^T) = p$.

The vector of orthogonal data from Eq. (2.9) is $\tilde{\mathbf{y}}$, and the corresponding $\tilde{\mathbf{y}}^T$ matrix is such that

$$\tilde{\mathbf{y}}^T = \begin{pmatrix} \mathbf{P}^T\mathbf{X}^T \\ \mathbf{H}^T \end{pmatrix} \tag{3.51}$$

which is also equivalent to

$$\begin{pmatrix} E(\tilde{\mathbf{y}}) \\ \mathbf{0} \end{pmatrix} = \begin{pmatrix} \mathbf{P}^T\mathbf{X}^T \\ \mathbf{H}^T \end{pmatrix} \tilde{\beta} = \begin{pmatrix} \mathbf{P}^T\mathbf{X}^T \\ \mathbf{0} \end{pmatrix} \beta \tag{3.52}$$

Multiplying both sides by $\tilde{\mathbf{y}}^T$, we obtain

$$\begin{pmatrix} \mathbf{XP} & \mathbf{H} \end{pmatrix} \begin{pmatrix} \mathbf{P}^T\mathbf{X}^T \\ \mathbf{H}^T \end{pmatrix} \tilde{\beta} = \begin{pmatrix} \mathbf{XP} & \mathbf{H} \end{pmatrix} \begin{pmatrix} \mathbf{P}^T\mathbf{X}^T \\ \mathbf{0} \end{pmatrix} \beta \tag{3.53}$$

Using the fact that $\mathbf{PP}^T = \mathbf{K}_f^{-1}$ and expanding Eq. (3.53), we have

$$(\mathbf{XK}_f^{-1}\mathbf{X}^T + \mathbf{HH}^T)\tilde{\beta} = \mathbf{XK}_f^{-1}\mathbf{X}^T\beta \tag{3.54}$$

Since $\tilde{\mathbf{y}}^T$ is of rank p, the vector of parameters $\tilde{\beta}$ is then

$$\tilde{\beta} = (\mathbf{XK}_f^{-1}\mathbf{X}^T + \mathbf{HH}^T)^{-1}\mathbf{XK}_f^{-1}\mathbf{X}^T\beta \tag{3.55}$$

Because $E(\mathbf{y}) = \mathbf{X}^T\beta$, it implies that $\widehat{\beta}$ has the unbiased estimate

$$\widehat{\tilde{\beta}} = (\mathbf{XK}_f^{-1}\mathbf{X}^T + \mathbf{HH}^T)^{-1}\mathbf{XK}_f^{-1}\mathbf{y} \tag{3.56}$$

Using Eqs. (3.42) and (3.56), the final form of $SS_e(\mathbf{y}, \beta)$ is then

$$SS_e(\mathbf{y}, \beta) = (\mathbf{Qy})^T\mathbf{K}_f^{-1}(\mathbf{Qy}) \tag{3.57}$$

where

$$\mathbf{Q} = \mathbf{I} - \mathbf{X}^T(\mathbf{XK}_f^{-1}\mathbf{X}^T + \mathbf{HH}^T)^{-1}\mathbf{XK}_f^{-1} \tag{3.58}$$

The detector is implemented in a similar manner as in Section 3.5. The only additional operation is the determination of the contrast variance using Eq. (3.48) to take into account the correlated nature of the corrupting noise. Fig. 3.16 shows the result using the present detector on the image shown in Fig. 3.15.

3.7 Trajectory detection

3.7.1 Unidirectional detectors

In the present section we deal with the detection of trajectories present in a noisy scene. The problem is similar to that of line detection in that the present procedure can also be applied to the extraction of lines. We specifically deal with incomplete

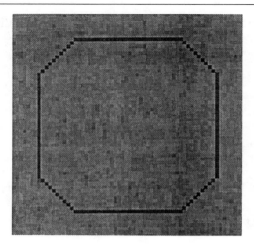

Figure 3.15. Correlated noisy version of Fig. 3.12(a). The horizontal and vertical correlation coefficients $a = b = 0.4(\sigma = 8)$.

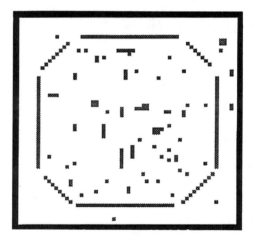

Figure 3.16. Output of detector designed for correlated noise.

block-based ANOVA designs. We start with the unidirectional case using the SBIB design. Then we extend the analysis to the Youden squares,[1] which allow the simultaneous detection of any two directions separated by 90°.

Methodology

A trajectory is defined as a continuous region of high intensity pixels surrounded by a region of low intensity pixels on each side. It is characterized by two properties:

---

[1]Youden squares are related to the SBIB designs as GLS designs are related to LS designs.

(1) The block nature of the pixel gray level distribution, and (2) the strong contrast of trajectory elements.

Those elements containing a trajectory tend to be adjacent to one another: their gray levels or textures tend to be similar when compared to the surrounding elements on both sides. The gray level distribution changes drastically on both sides of the edges.

The block nature of the design is such that the block effects associated with strong contrasts are indicative of a trajectory. Consequently, one can use a design that allows the inclusion of blocks in the description of the collected data. A complete block design is appropriate to the present characterization of the trajectory problem: three parameters can be used in the model. These are the column and row position information that determines the planar location and the gray level distribution that describes the relative contrast.

The requirement that all treatments appear once in each block of a completely randomized design makes the size of the design too large for practical implementation if there is a large number of treatments. In addition, a window with large size has a rather poor resolution capability. An alternate possibility is in the use of a balanced incomplete design for which the blocks are of the same size and each treatment appears the same number of times. The necessary conditions for the existence of a BIB design are given in Section 2.5.

The application of BIB designs in image processing problems in general and trajectory detection in particular permits the comparison of a large number of treatments within relatively small blocks. This reduces the information loss due to quantization of treatment levels. A particular class of BIB designs, referred to in the literature as the symmetric BIB (SBIB), is one for which $b = t$, that is, the number of blocks is equal to the number of treatments in the design. These designs have proven to be very useful in trajectory detection.[20, 21] Although the construction of SBIB designs is beyond the scope of this section, it is sufficient to say that SBIB designs with

$$
\begin{aligned}
v = b &= 4t + 3 \\
\rho = k &= 2t + 1 \\
\lambda &= t
\end{aligned}
\tag{3.59}
$$

can always be constructed when $4t + 3$ is a prime or a prime power.

The analysis is carried out by considering the hypotheses H_a : all $\alpha_i = 0$ and H_b : all $\beta_j = 0$. The F-statistics used in this section are given by Eqs. (2.109) and (2.110). In the present case we use exclusively F_a for the test of block effects to determine whether there is a trajectory in the window.

Because the number of blocks is larger than the number of units in each block, the SBIB design has the shape of a rectangle. As an additional result, the analysis yields the adjusted block means, which are used as the unbiased estimate of the true block means. Chang and Kurz[21] have shown that the hypothesis that the window contains a trajectory is accepted if the following two conditions are satisfied: (1) There exists significant block effects, and (2) the adjusted block means behave like a pulse function.

   If the first condition is satisfied then one can state that there are edges present. If further, the block effects are qualified in a comparison test setting, then only those edges with strong contrast are identified.

   To differentiate between a possible edge and a trajectory, one has to use the adjusted block means in a threshold test setting that ultimately decides on the exact nature of the feature in the window.

Detection of feature

The detection is implemented using Eq. (2.109). The hypothesis that all block effects are equal is rejected whenever F_a exceeds the threshold corresponding to a level of the test α. Note that the specific nature of the feature in the window is not yet determined at this stage.

Identification and location of trajectories

a-Identification by block means   The shape of the gray level distribution, that is, the pulse-like distribution, is determined using a thresholding test of the adjusted block means. The unbiased estimate of the mean of the qth block is defined by

$$\widehat{b}_q = y_{...} + \frac{t}{\lambda t}\left(y_{.q.} - \sum_{i=1}^{t} n_{iq}y_{i..}\right) \tag{3.60}$$

where

$y_{...}$ is the overall arithmetic mean

$y_{.q.}$ is the sum of the measurements from a particular gray level

$y_{i..}$ is the arithmetic mean of the measurements in ith block

The overall mean b_m is defined as

$$b_m = \frac{1}{b}\sum_{i=1}^{b} \widehat{b}_i \tag{3.61}$$

   Using Eqs. (3.60) and (3.61), b_m and the block means $\widehat{b}_q; q = 1, 2, \ldots, m$, are used to form the basis of the threshold testing scheme. Let H_i be the threshold function that compares each individual block mean with b_m. We can write

$$H_i = U(\bar{b}_i - b_m) \quad \begin{matrix} < & \textit{Put 0 in block position} \\ \\ > & \textit{Put 1 in block position} \end{matrix} \tag{3.62}$$

After the thresholding is terminated, the result is a series (n_i) of 1 and 0 that can be used to determine whether we are in the presence of an edge or a trajectory. Using a sign test of the form

$$S_i = \begin{cases} + \text{ if } n_i \neq n_{i+1} & i = 1, 2, \ldots, (b-1) \\ - \text{ if } n_i = n_{i+1} \end{cases} \tag{3.63}$$

on the series (n_i), we are able to determine whether there is a change of value between two adjacent blocks. Because we are interested in detecting a trajectory, we need to have two '+' signs in the present encoding as shown in Fig. 3.17. Any other possibility is rejected, and the trajectory location is obtained by considering the location of the '+' signs in the window.

$$
\begin{array}{ccccccc}
o & o & x & x & x & o & o \\
o & o & x & x & x & o & o \\
o & o & x & x & x & o & o \\
\end{array}
$$

$$
\begin{array}{lccccccc}
N_p = & 0 & 0 & 1 & 1 & 1 & 0 & 0 \\
S_p = & - & + & - & - & - & + & - \\
\end{array}
$$

Figure 3.17. Sign test for a trajectory located in the center of the mask, $d = 3$.

b-Identification by comparisons　An alternate technique for locating a trajectory and identifying its contrast is based on the multicomparison methodology. The main element here is the block mean (B_i). Recall that a contrast function can be defined and is such that $\psi = \sum_{i=1}^{b} c_i B_i$ where the coefficients c_is are known constants subject to the zero-sum condition. As mentioned previously, when the F-test is rejected, the location of the specific mean responsible for the rejection is possible; that is, there exists a contrast function that is significantly different from zero. In a hypothesis setting, we have

$$
\begin{aligned}
H_o : & \ \psi = 0 \\
H_i : & \ \psi \neq 0
\end{aligned} \tag{3.64}
$$

In this section we deal exclusively with the one-dimensional contrast case. For a trajectory of width d, the set of coefficients is such that[2]

$$c_i = \begin{cases} 1 & i = (b-d)/2 + k \quad k = 1, 2, \ldots, d. \\ -d/(b-d) & \text{otherwise} \end{cases} \tag{3.65}$$

where the trajectory is assumed to be located in the center of the window as shown

---

[2]We assume here that b and d are both odd numbers.

in Fig. 3.18. As an example, for a 3×7 window, the contrast function is

$$\psi = (B_3 + B_4 + B_5) - \frac{3}{4}(B_1 + B_2 + B_6 + B_7) \tag{3.66}$$

```
o o x x x o o
o o x x x o o
o o x x x o o
```

Figure 3.18. Trajectory located in the center of the mask, $d = 3$.

As a result, it is possible to determine whether the blocks B_1, B_2, and B_3 have a strong contrast as compared to the rest of the blocks in the window.

Note that if the trajectory width d is fixed a priori, the window size should be chosen accordingly. In addition, it is possible to use additional contrast functions besides the one used for the center location. However, because multiple comparisons tend to increase the number of type II errors, it is in general preferable to use a single contrast function.

3.7.2 Bidirectional trajectory detectors

While a SBIB-based detector can be used in detecting trajectories in a single direction, a Youden square (YS) design is more effective in that it permits the detection of single as well as dual directions (with $90°$ separation). This is accomplished by using not only significant block effects but also significant unit effects as well. In contrast to the SBIB design, the YS has an additional parameter referred to as the "unit effect," with the restriction that every treatment occurs exactly once in each unit.

By adding the unit effect δ_l to the SBIB model, we obtain the YS parameterization, that is,

$$\Omega : \begin{cases} y_{ijl} = \mu + \alpha_i + \beta_j + \delta_l + e_{ijl} \\ i = 1, 2, \ldots, t; j = 1, 2, \ldots, b; l = 1, 2, \ldots, k. \\ \sum_{i=1}^{t} \alpha_i = 0; \sum_{j=1}^{b} \beta_j = 0; \sum_{l=1}^{k} \delta_l = 0. \\ (e_{ijl}) \; is \; N(0, \sigma^2 \mathbf{I}) \end{cases} \tag{3.67}$$

We assume that interactions between effects are nonexistent.

The analysis is similar to that of the SBIB design but with the addition of the sum of squares due to the effect δ_l. The F-statistics of treatments, blocks, and unit are, respectively

$$F_t = \frac{SS_t(\mathbf{y}, \beta)/(t-1)}{SS_e(\mathbf{y}, \beta)/(bk - t - b - k + 2)} \tag{3.68}$$

$$F_b = \frac{SS_b(\mathbf{y}, \beta)/(b-1)}{SS_e(\mathbf{y}, \beta)/(bk - t - b - k + 2)} \tag{3.69}$$

and

$$F_k = \frac{SS_k(\mathbf{y}, \beta)/(k-1)}{SS_e(\mathbf{y}, \beta)/(bk - t - b - k + 2)} \qquad (3.70)$$

where the respective sum of squares components are

$$SS_t(\mathbf{y}, \beta) = \frac{k}{\lambda t} \sum_{i=1}^{t} \left(y_{i..} - 1/k \sum_{j=1}^{b} n_{ij} y_{.j.} \right)^2 \qquad (3.71)$$

$$SS_b(\mathbf{y}, \beta) = \frac{\rho}{\lambda t} \sum_{i=1}^{t} \left(y_{.j.} - 1/\rho \sum_{j=1}^{b} n_{ij} y_{i..} \right)^2 \qquad (3.72)$$

and

$$SS_k(\mathbf{y}, \beta) = \frac{1}{b} \sum_{i=1}^{t} y_{..l}^2 - 1/bk \, y_{...}^2 \qquad (3.73)$$

and the error sum of squares under the alternative is derived in a straightforward way.

In its application to trajectory detection, the Youden square detects significant block effects as well as significant unit effects. This enables the detection of lines in the 0° and 90° directions.

Identification and location of trajectories
Both block and unit effects are used in the identification of trajectories using the thresholding scheme of Section 3.7.1. The block-based approach is directly applicable, but the problem is somewhat more complicated for the unit-based approach since by design the block size is very small and contains very few units to yield any meaningful results. Recall that the common window size that we use is the 3×7 where the block size is 3, which yields a poor resolution for the detection of the pulse-like shape distribution.

Chang and Kurz[21] discuss some interesting schemes for solving the small block size. In the first scheme they use the configuration of Fig. 3.19 where additional YS designs are placed in parallel to the original YS design. In this case one can use the thresholding scheme. In the second scheme an additional YS design is placed in parallel with the original YS design. To determine the side where the design is placed, a first run of the thresholding scheme is performed by using the unit effect against the overall unit mean. The resulting series of 1 determines the side where the YS design is to be placed. A second thresholding is run on the additional YS. By concatenating the resulting series (n_i) with the original, one is able to determine whether there is a trajectory in the horizontal direction. If the result is positive, a sign test will determine its exact location.

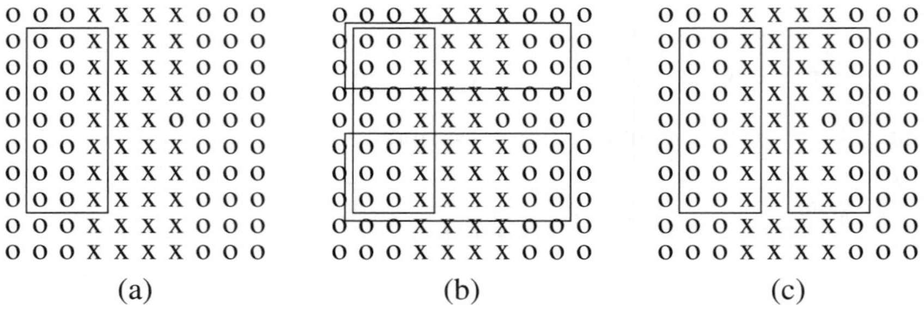

```
O O O X X X X O O O        O O O X X X X O O O        O O O X X X X O O O
O O O X X X X O O O        O O O X X X X O O O        O O O X X X X O O O
O O O X X X X O O O        O O O X X X X O O O        O O O X X X X O O O
O O O X X X X O O O        O O O X X X X O O O        O O O X X X X O O O
O O O X X O O O O O        O O O X X O O O O O        O O O X X O O O O O
O O O X X X X O O O        O O O X X X X O O O        O O O X X X X O O O
O O O X X X X O O O        O O O X X X X O O O        O O O X X X X O O O
O O O X X X X O O O        O O O X X X X O O O        O O O X X X X O O O
O O O X X X X O O O        O O O X X X X O O O        O O O X X X X O O O
O O O X X X X O O O        O O O X X X X O O O        O O O X X X X O O O
```

(a)   (b)   (c)

Figure 3.19. (a) Youden square designs, (b) perpendicular scheme, (c) parallel scheme.

3.8 Multidirectional adaptive detectors

When there are structured backgrounds in the image, the number of false alarms tends to rise significantly in the vicinity of the structure. The procedures developed before can be modified to handle these potential problems. Habestroh[16] describes a number of algorithms using the GLS model.

The idea here is to restrict the alternative to specific features of interest. As we saw in Section 3.5, the UMP conditions are not always met, and for that reason tests of shape for lines in various directions were introduced. The choice of the alternative is dependent on the nature of the problem at hand. If the alternative is chosen, for example, to reflect the occurrence of a line through the center of the window, the hypothesis-alternative pair can be expressed as

$$H_m : \text{ all } \theta_{m,i} \text{ are equal} \quad i = 1, 2, \ldots, 5$$
$$K_m : \theta_{m,1} = \theta_{m,2} = \theta_{m,4} = \theta_{m,5} \neq \theta_{m,3} \tag{3.74}$$

for a particular treatment group m. Note that one could have chosen a treatment level other than $\theta_{m,3}$. However, such a choice is inappropriate in the present case.

The importance of such a modification is in the reduction of the dimensionality of the hypothesis-alternative space from five to two levels when we assume that the original treatment formulation has five levels. Because this approach reflects a similar idea to that of the shape test, it becomes unnecessary to carry out a shape test. The problem can now be reformulated by defining new treatment effects for the groups of interest, which here can be written as

$$\phi_{m,1} = \theta_{m,3}$$
$$\phi_{m,2} = \theta_{m,n} \quad n \neq 3 \tag{3.75}$$

Note that at this stage we do not assume any specific ANOVA model. Recalling the general form of the linear model, any observation can be written as

$$y_{ij} = \mu + \phi_{1,2-\delta(i')} + \theta_{2,j} + \ldots + \theta_{I,j} + e_{ij} \tag{3.76}$$

where we assume there are I treatments in the design and the primary group[3] is θ_1. Here δ is the Kronecker delta and $i' = i - 3$. It is important to stress that Eq. (3.76) is valid as long as we consider a centrally located line. As such, any background structure that occurs in the same group θ_m as the line will make the assumptions inappropriate. Habestroh[16] discusses an alternate approach to deal with this problem.

The estimates of the primary treatment groups as defined by Eq. (3.75) are

$$\widehat{\phi}_{1,1} = y_{3.} - y_{..} \tag{3.77}$$

and

$$\widehat{\phi}_{1,2} = (1/4)[y_{1.} + y_{2.} + y_{4.} + y_{5.}] - y_{..} \tag{3.78}$$

or in terms of the old treatments by

$$\widehat{\phi}_{1,1} = \widehat{\theta}_{m,3} \tag{3.79}$$

and

$$\widehat{\phi}_{1,2} = (1/4)[\widehat{\theta}_{m,1} + \widehat{\theta}_{m,2} + \widehat{\theta}_{m,4} + \widehat{\theta}_{m,5}] \tag{3.80}$$

The numerator of the F-statistic, that is, BSS_m for the ϕ_m treatment group, assuming a GLS design, is

$$BSS_{\phi_m} = 5\widehat{\phi}_{m,1}^2 + 20\widehat{\phi}_{m,2}^2 \tag{3.81}$$

Recalling the side condition on the old treatment θ_m, which in terms of the new treatments ϕ_m is $\phi_{m,1} + 4\phi_{m,2} = 0$, Eq. (3.81) can be written as

$$BSS_{\phi_m} = 6.25\widehat{\phi}_{m,1}^2 = 6.25(y_{3.} - y_{..})^2 \tag{3.82}$$

The sum BSS_{ϕ_m} can be partitioned into orthogonal components[2] by using "component sum of squares" CSS, which are based on the contrast functions of Chapter 2.

Let $\psi = \sum_{i=1}^{m} c_i \theta_{m,i}$ with $\sum_{i=1}^{m} c_i = 0$, the CSS due to ψ is equal to $\psi^2 / \sum_{i=1}^{m} c_i^2$. Because there are m levels and given the orthogonality condition, there are at most $m - 1$ orthogonal contrast functions that can be used to obtain $(m - 1)$ CSS sums. Returning to our problem, it is obvious that the obtained CSS_{ϕ_m} is identically equal to BSS_{ψ_m} since we have only two levels.

Until now, we assumed that only the primary treatment levels can be reduced to two levels while the remaining secondary groups were allowed to maintain five levels. A natural question that arises is whether under certain conditions all treatments, primary as well as secondary, can be simultaneously reduced to two

---

[3]A primary group is the treatment group on which the hypothesis is based, while the secondary groups refer to the remaining treatments.

levels. The answer lies in part in the prior knowledge of the scene, especially about potential structures and lines in the image. If one knows that secondary lines are not centrally located, one can conclude that only the primary group is to be reduced in size. On the other hand, if it is known that all potential lines pass through the center of the window, then all treatments groups could be reduced. A detector based on the GLS model, assuming that all treatment groups are reduced in size, has the form

$$F_d = \max \quad 8 \frac{CSS_m}{WSS} U(s_m) \quad m = 1, 2, 3, 4. \tag{3.83}$$

with

$$T_H = F_{1,8,\alpha'} \tag{3.84}$$

and

$$\alpha' = 2[1 - (1 - \alpha)^{1/4}] \tag{3.85}$$

Here we have two levels for each treatment, hence the degree of freedom for the CSS.

As suggested in reference [16], Eq. (3.83) can be modified to reduce the number of false alarms in the vicinity of edges that arise from intensity discontinuity in the image.

In this case, two CSSs, which are used to test separately the contrast of the central mean with respect to means on each side, are formed. The contrasts are given by

$$\psi_L = 2\theta_{1,3} - (\theta_{1,1} + \theta_{1,2}) \quad \text{with } c_1 = (-1 \quad -1 \quad 2 \quad 0 \quad 0) \tag{3.86}$$

and

$$\psi_R = 2\theta_{1,3} - (\theta_{1,4} + \theta_{1,5}) \quad \text{with } c_2 = (0 \quad 0 \quad 2 \quad -1 \quad -1) \tag{3.87}$$

Let $s* = \min(\psi_L, \psi_R)$. The detector structure is then given by

$$F_d = \max \quad 8 \frac{CSS_m^*}{WSS} U(s_m^*) \quad m = 1, 2, 3, 4. \tag{3.88}$$

with

$$\alpha' = (1/.402)[1 - (1 - \alpha)^{1/4}] \tag{3.89}$$

The contrast functions selected for this detector are not optimum in the sense that they do not satisfy the orthogonality condition on contrasts (see Chapter 6), that is, $\sum_{i=1}^{m} c_{i1} c_{i2} \neq 0$. An appropriate set of functions is given by the following set of coefficients

$$
\begin{aligned}
c_1 &= (-1 \quad -1 \quad 2 \quad 0 \quad 0 \;) \\
c_2 &= (\; 0 \quad\;\;\; 0 \quad 0 \quad 1 \quad -1) \\
c_3 &= (\; 1 \quad -1 \quad 0 \quad 0 \quad 0 \;)
\end{aligned} \tag{3.90}
$$

where c_2 and c_3 are used to check, separately, whether off-center adjacent means are equal. Let $w* = \min(\psi_1, \psi_2, \psi_3)$. The detector structure is then given by

$$
F_d = \max 8 \, \frac{CSS_m^*}{WSS} U(w_m^*) \quad m = 1, 2, 3, 4. \tag{3.91}
$$

with

$$
\alpha' = (1/.402)[1 - (1 - \alpha)^{1/4}] \tag{3.92}
$$

3.9 Concluding remarks

In this chapter we present several procedures for detecting lines and trajectories. We first start with the unidirectional problem, where we use the one-way based ANOVA model in developing the detector structure. We then extend the problem to include the detection in orientations separated by 90°, by using the two-way based ANOVA model. The advantage in using this detector over the one-way based detector is highlighted.

Next we present a multidirectional detector that enables the detection in four directions, that is, 0°, 45°, 90°, and 135°. Then, a version of the detector modified for correlated noise environments is presented.

We then introduce contrast-based detectors, which solve the thickening problems that arise when using the F-statistics in the detection without restricting the number of alternatives.

Trajectory detection is handled by using incomplete designs, in this case the SBIB and YS designs, which perform well in unidirectional and bidirectional detection, respectively.

Finally, the adaptive GLS structure is discussed. It is shown that by chosing the number of alternatives in the detection, selected features can be detected.

4

Edge detection

4.1 Introductory remarks

In the preceding chapter we introduced several techniques based on localized mask operations for detecting local features such as lines, broken lines, and trajectories. The results obtained in both uncorrelated and correlated noisy environments lead us to expect that similar techniques can be developed for the detection of a related class of features, namely edges. To understand the similarity as well as the difference between line and edge detection problems, consider Fig. 4.1. In the ideal case, assuming two gray levels without noise, an edge appears as a step-like change as one moves from one region to the other, or the change of the gray level is assumed to be abrupt as the border is crossed. Note that both regions are adjacent as well as relatively large.

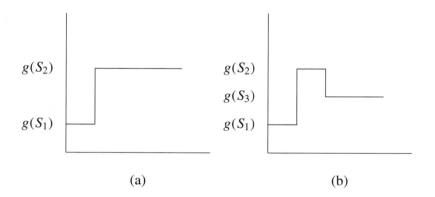

$$(a) \qquad\qquad\qquad (b)$$

Figure 4.1. Step-like distribution of edge pattern. (a) Ideal edge step, (b) line-edge crossing.

On the other hand, as explained in Section 3.1, lines appear as having a pulse-like distribution with narrow width. As such, we have two crossings between gray levels. Sometimes, there is also the possibility of having an edge combined with a line, which happens when the two regions have almost the same gray level.

We discuss the methodology for detecting edges in Section 4.2. Specifically, we consider a simple detector used in detecting edges oriented in one direction. A bidirectional detector is then considered. Subsequently, the theory is extended to include the multidirectional detectors based, respectively, on the Latin square and Græco-Latin square masks. The detection in correlated environments is undertaken by considering a Latin-square-based detector for which the corrupting noise is assumed to be a first-order Markov linear process. Finally, we consider the reconstruction of edges by simple mask operations.

4.2 Edge detection methodology

In the design of an efficient edge detector, one has to take into account two constraints. First, the data set is corrupted by noise, often of unspecified nature, which results in blurred edges and false features. Second, an edge is sometimes associated with a line transition that further complicates the edge detection process. As such, it is advantageous to use a statistical approach that would provide both robustness and discrimination of lines in the course of detecting possible edges.

Like the line problem of Chapter 3, the tests for statistical significance in this chapter are performed in two steps. The initial step is the formation of test statistics by parameter estimation given the particular nature of the ANOVA model under consideration. Then, hypothesis tests are applied to detect the presence or absence of features in particular directions.

Whenever the exact location of an edge is desired, a comparison test is performed on the data. The approach is based on the multicomparison tests introduced in Chapter 2.

In the development of edge extractors, we need to define precisely the edge characteristics in terms of the model that is used in the representation of the collected data. The following remarks are useful when using a mask with dimensions $m \times n$ over an image containing two adjacent regions S_1 and S_2 with two gray level values denoted by $g(S_1)$ and $g(S_2)$, respectively.

(1) If the mask is entirely over either S_1 or S_2, then a hypothesis of the form H : all $\alpha_i = 0$, where α_i is a certain treatment from a particular ANOVA model, is accepted with a level of the test α. This is a direct result of the homogeneity of the mask.

(2) If the mask is in the vicinity of the transition between S_1 and S_2, then H is rejected. The main factor affecting the rejection is the heterogeneity of the mask. This is due to the presence of elements having both gray levels $g(S_1)$ and $g(S_2)$. Consequently, we can declare with a significance

level α that a certain feature is present within the mask, though we cannot specifically determine its nature.

(3)   If the hypothesis is rejected, one can use the same approach as developed in Chapter 3 for the detection of lines of various orientations. As such, with the present characterization, a line may cause the rejection of the hypothesis. Since that is not of any interest to us, we have to consider a further refinement by introducing tests based on the exact nature of the edges under consideration. Fig. 4.2 shows four types of possible edges. A possible test feature is of the form $\psi = c_1 \sum_i \alpha_i - c_2 \sum_j \alpha_j$ where c_1 and c_2 are constants that depend on the exact number of effects in regions S_1 and S_2, respectively.

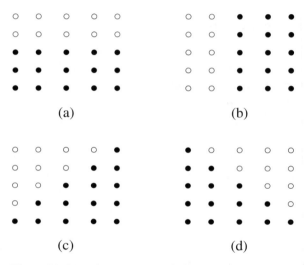

Figure 4.2. Edge structures. (a) Horizontal, (b) vertical, (c) diagonal 45°, (d) diagonal 135°.

4.3 Unidirectional edge detectors

The basic structure of a unidirectional edge detector is similar to the unidirectional line detector developed in Section 3.2. The two problems are essentially the same if one considers the F-test methodology. When the hypothesis is rejected, the cause is generally unknown, that is, it could have been the result of a line, edge, or any other feature contained within the localized mask. To provide a way to solve this ambiguity, we first develop the basic detector structure before moving to a more elaborate scheme that includes shape tests.

Assume that the feature of concern is an edge oriented vertically as shown in Fig. 4.2(b). Note that the approach can be readily adapted to the detection of horizontal edges. The data model is the basic one-way model with β_j designating vertical effects. The observations set is then characterized by

$$\Omega : \begin{cases} y_{ij} = \mu + \beta_j + e_{ij} & i = 1, 2, \ldots, m; j = 1, 2, \ldots, n. \\ \sum_{j=1}^{n} \beta_j = 0. \\ (e_{ij}) \text{ independent } N(0, \sigma^2 \mathbf{I}) \end{cases} \tag{4.1}$$

The estimates of μ and β_j, $j = 1, 2, \ldots, n$ are obtained by considering the usual least squares approach, in which case we obtain

$$\begin{aligned} \widehat{\mu} &= y_{..} \\ \beta_j &= y_{.j} - y_{..} \quad j = 1, 2, \ldots, n \end{aligned} \tag{4.2}$$

and the test of the hypothesis $H :$ all $\beta_j = 0$ is accomplished by comparing

$$F = \frac{m \sum_{j=1}^{n} (y_{.j} - y_{..})^2 / (n-1)}{\sum_{i=1}^{m} \sum_{j=1}^{n} (y_{ij} - y_{.j})^2 / n(m-1)} \tag{4.3}$$

to the threshold $F_{\alpha, n-1, n(m-1)}$ for a significance level α. Conceptually, the above test is simply the test of whether the columns are homogeneous in gray level. Used on images that incorporate lines in addition to edges, the one-way detector structure would result in a rather high level of false alarms. Therefore, it is natural to incorporate an additional test to remove this ambiguity. One such technique, related to the shape test and used by Behar[14] in edge-transition methodologies, is essentially based on localization of specific strings of pixel arrangement while moving the scanning mask over the image.

Consider the mask $\mathbf{Y}(i, j)$ in the vicinity of an edge. Assume that the gray levels $g(S_1)$ and $g(S_2)$ are represented by 0 and 1, respectively. Then, the possible strings are 00000, 00001, 00011, 00111, 01111, and 11111. The first string corresponds to the mask being entirely over S_1 while the last one corresponds to the mask over S_2. The strings in between these extremes correspond to the mask moved one column to the right. The strings that most likely correspond to an actual edge are 00011, 00111 since the first and last strings correspond to homogeneous regions. In addition, the remaining strings are mirror images, that is, 00001 is the negative of 01111, while 00011 is the complement of 00111. The possible contrasts (see Chapter 2 for definition) are then

$$\psi_1 = 3(\beta_4 + \beta_5) - 2(\beta_1 + \beta_2 + \beta_3) \tag{4.4}$$

and

$$\psi_2 = 2(\beta_3 + \beta_4 + \beta_5) - 3(\beta_1 + \beta_2) \tag{4.5}$$

Based on numerous simulations on test images,[14] ψ_2 is selected as the additional test statistic. By replacing the effects β_j by their estimates, the corresponding contrast estimate is then

$$\widehat{\psi}_2 = 2(\widehat{\beta}_3 + \widehat{\beta}_4 + \widehat{\beta}_5) - 3(\widehat{\beta}_1 + \widehat{\beta}_2) \tag{4.6}$$

Using results from Chapter 2, the threshold in a confidence interval setting is

$$|\widehat{\psi}_2| \overset{<}{\underset{>}{}} (Ss^2 F_{\alpha,1,n-r})^{1/2} \tag{4.7}$$

where in the independent data case S is given by Eq. (2.112). In terms of the effects weights, the mask values are as shown in Fig. 4.3.

−3	−3	2	2	2
−3	−3	2	2	2
−3	−3	2	2	2
−3	−3	2	2	2
−3	−3	2	2	2

Figure 4.3. Mask values for the shape test. Unidirectional detector.

4.4 Bidirectional edge detectors

When considering edges oriented in a single direction, one can use the previous detector structure. The parameters, either row or column effects, can be adjusted to reflect horizontal or vertical edges. Likewise, the shape test should be chosen accordingly. If there is a possibility of having both types of edges present in the image, it is more appropriate to use a bidirectional edge detector. The design is similar to the line detector, but the shape test is modified to reflect the difference between line and edge features. Consequently, the detector structure is similar to the one developed in Section 3.3. With the hypotheses H_a : all $\alpha_i = 0$ and H_b : all $\beta_j = 0$, the test statistics are

$$F_a = \frac{n \sum_{i=1}^{m}(y_{i.} - y_{..})^2/(m-1)}{\sum_{i=1}^{m}\sum_{j=1}^{n}(y_{ij} - y_{i.})^2/(n-1)(m-1)} \tag{4.8}$$

and

$$F_b = \frac{m \sum_{j=1}^{n}(y_{.j} - y_{..})^2/(n-1)}{\sum_{i=1}^{m}\sum_{j=1}^{n}(y_{ij} - y_{.j})^2/(n-1)(m-1)} \tag{4.9}$$

The thresholds for rejection of the hypotheses are $F_{\alpha,m-1,(m-1)(n-1)}$ and $F_{\alpha,n-1,(m-1)(n-1)}$, respectively. The shape test in the bidirectional case is

$$s(\gamma) = 2(\gamma_3 + \gamma_4 + \gamma_5) - 3(\gamma_1 + \gamma_2) \tag{4.10}$$

where γ denotes row or column effects depending on the edge under consideration. By replacing $\gamma_i, i = 1, 2, \ldots, 5$ by their estimates in Eq. (4.10), we obtain the shape estimate

$$s(\hat{\gamma}) = 2(\hat{\gamma}_3 + \hat{\gamma}_4 + \hat{\gamma}_5) - 3(\hat{\gamma}_1 + \hat{\gamma}_2) \tag{4.11}$$

In terms of the effects values, the shape masks have the values shown in Fig. 4.4.

-3	-3	2	2	2		-3	-3	-3	-3	-3
-3	-3	2	2	2		-3	-3	-3	-3	-3
-3	-3	2	2	2		2	2	2	2	2
-3	-3	2	2	2		2	2	2	2	2
-3	-3	2	2	2		2	2	2	2	2

Figure 4.4. Mask values for shape tests. Bidirectional detectors.

4.5 Multidirectional edge detectors

In terms of the number of simultaneous edges that can be detected, the bidirectional detector presents a notable improvement over the unidirectional detector. However, it does not approximate well the probable realization of edges in actual situations. It is impossible, for instance, to account for diagonally oriented edges with the present structure. What we need is a structure capable of detecting the four possible edge orientations, namely, the horizontal, vertical, and diagonals (45° and 135°).

The LS and GLS mask-based detectors are the most commonly used in similar types of problems. Although the GLS detector includes all four directions, the first design to be proposed was based on the Latin square.[11] The main motivation for this selection is the possibility of modeling vertical and horizontal edges using a block approach, that is, vertical and horizontal blocks of pixel elements. The third parameter in the design is used to model gray level homogeneity within the mask. Consequently, the Latin square model was used for only two directions in addition to the test of sharpness of the eventual edge. To see that more clearly, assume that the data is of a block nature, which implies that adjacent pixels tend to have the same gray level and by extension the same texture. For this modeling, the row and column classifications are viewed as blocks of pixels. As a result, the hypotheses account for differences between blocks, vertical and horizontal, as well as differences among selected pixels of different gray levels between blocks, regardless of their assignment within the blocks.

4.5.1 Latin square-based detector

As we already pointed out in Section 2.5.1, the Latin letters in the Latin square mask designate treatments, while the rows and columns refer to the horizontal and vertical blocks. Fig. 4.5 shows a typical 5×5 randomized Latin square layout. Each gray level occurs once, and only once, in each row and in each column of the layout. Each block of pixels is enclosed in a box to show that it consists of multiple elements. Note that the layout is called "randomized" because the arrangement of the Latin letters does not follow any order. In general, before testing for effects one must proceed with a permutation of the rows and columns to ensure that any bias is removed. As pointed out in Kadar and Kurz,[11] one does not know a priori the distribution of gray levels within the Latin square mask, so one may assume that they were randomized beforehand.

```
A   B   C   D   E
B   E   A   C   D
C   D   B   E   A
D   C   E   A   B
E   A   D   B   C
```

Figure 4.5. A randomized 5×5 Latin square.

A natural question to be raised at this point is what advantages one gains by using a Latin square model rather than the more conventional two-way based detector. Clearly, there is the additional test for gray level homogeneity, which can be used to check whether there are any significant gray level differences, a clear sign that there is a well-defined edge. By using this information one would decide whether to proceed to the test of presence of vertical and horizontal edges. The amount of computation is thus drastically reduced because we do not need to carry out all three tests.

However, this is a somewhat minor improvement. A more clever use of the Latin square is in the use of gray levels as a way to detect diagonally oriented edges. To see that, one has to consider all possible arrangements for the Latin square designs of size $m \times n$. For example, using a 5×5 design, one can always arrange the Latin letters so as to obtain features of particular interest. An example of such arrangement is shown in Fig. 4.6(a). Commonly referred to as a "systematic square," the present arrangement is used to detect gray level edges parallel to diagonals oriented in the $45°$ direction. Another form of this specific arrangement, shown in Fig. 4.6(b), allows detection in the $135°$ direction.

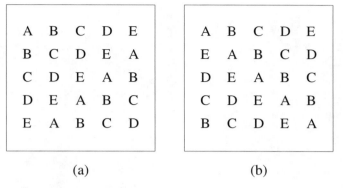

Figure 4.6. Systematic 5×5 Latin squares.

For the present characterization, we assume the representation of Fig. 4.6(a). The observations y_{ijk}, where the triplets (i, j, k) take on only m^2 values out of the m^3 possible values, are characterized by the following model

$$\Omega : \begin{cases} y_{ijk} = \mu + \alpha_i + \beta_j + \tau_k + e_{ij} & (i, j, k) \in S \\ \sum_{i=1}^{m} \alpha_i = 0; \sum_{j=1}^{m} \beta_j = 0; \sum_{k=1}^{m} \tau_k = 0. \\ (e_{ij}) \text{ independent } N(0, \sigma^2 \mathbf{I}) \end{cases} \qquad (4.12)$$

under the assumption of additivity of the parameters, which are assumed for the time being to be fixed. The hypotheses of interest are as usual

$$\begin{array}{lll} H_a : \text{all } \alpha_i = 0 & \text{rows are homogeneous in pixel intensity} \\ H_b : \text{all } \beta_j = 0 & \text{columns are homogeneous in pixel intensity} & (4.13) \\ H_c : \text{all } \tau_k = 0 & \text{diagonals are homogeneous in pixel intensity} \end{array}$$

Unlike the case considered by Kadar and Kurz,[11] hypothesis H_c is used mainly for the modeling of diagonal edges. To find the test statistics corresponding to each hypothesis, we derive the least squares estimates of the parameters. These are the solutions to the normal equations (see Section 2.5.1). Hence,

$$\begin{array}{lll} \widehat{\mu} &= y_{...} \\ \widehat{\alpha}_i &= y_{i..} - y_{...} & i = 1, 2, \ldots, m \\ \widehat{\beta}_j &= y_{.j.} - y_{...} & j = 1, 2, \ldots, m \\ \widehat{\tau}_k &= y_{..k} - y_{...} & k = 1, 2, \ldots, m \end{array} \qquad (4.14)$$

The F-tests for testing H_a, H_b, and H_c at a significance level α are then

$$F_a = \frac{(SS_a(\mathbf{y}, \beta) - SS_e(\mathbf{y}, \beta))/(m - 1)}{SS_e(\mathbf{y}, \beta)/(m - 1)(m - 2)} \qquad (4.15)$$

$$F_b = \frac{(SS_b(\mathbf{y}, \beta) - SS_e(\mathbf{y}, \beta))/(m - 1)}{SS_e(\mathbf{y}, \beta)/(m - 1)(m - 2)} \qquad (4.16)$$

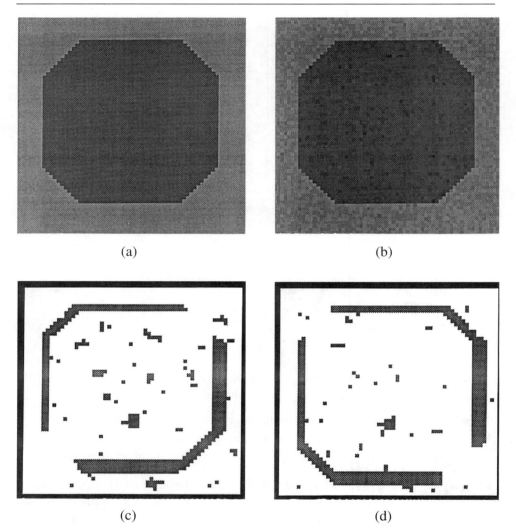

(a)   (b)

(c)   (d)

Figure 4.7. Edge image with four edge orientations. (a) Clean image, (b) noisy edge image, $\sigma = 10$ (SNR $\simeq 10$ dB), (c) output of Latin square detector (designed for $\theta = 0°$, $\theta = 90°$, and $\theta = 45°$), (d) same Latin square detector (designed for $\theta = 0°$, $\theta = 90°$, and $\theta = 45°$).

and

$$F_c = \frac{(SS_c(\mathbf{y}, \boldsymbol{\beta}) - SS_e(\mathbf{y}, \boldsymbol{\beta}))/(m - 1)}{SS_e(\mathbf{y}, \boldsymbol{\beta})/(m - 1)(m - 2)} \tag{4.17}$$

with the threshold given by $F_{\alpha,(m-1),(m-1)(m-2)}$, and the sum of squares by Eqs. (2.58)–(2.61). Fig. 4.7(c) shows the result of applying the Latin-square-based detector to the input image in Fig. 4.7(a). Notice that the detector structure indicates the presence of edges within the mask but is not able to locate it. Consequently, the edge transition region is thickened. As a result, a shape test similar to that of Section 4.4 should be used.

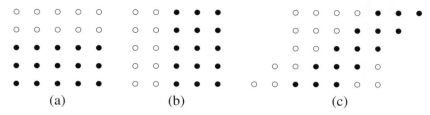

Figure 4.8. (a) Row contrast, (b) column contrast, (c) diagonal contrast.

Referring to Fig. 4.8(a), we can define a row contrast as

$$s_R(\alpha) = 2(\alpha_3 + \alpha_4 + \alpha_5) - 3(\alpha_1 + \alpha_2) \tag{4.18}$$

and its estimate is

$$s_R(\widehat{\alpha}) = 2(\widehat{\alpha}_3 + \widehat{\alpha}_4 + \widehat{\alpha}_5) - 3(\widehat{\alpha}_1 + \widehat{\alpha}_2) \tag{4.19}$$

Similarly, we can define a column contrast

$$s_C(\beta) = 2(\beta_3 + \beta_4 + \beta_5) - 3(\beta_1 + \beta_2) \tag{4.20}$$

For the shape tests, the estimate of the effects are the usual least squares estimates. Turning to the diagonal mask, we need to modify the definition of the contrast in a manner similar to Section 3.6 in the GLS structure case. Hence, we have

$$s_D(\widehat{\gamma}) = 2(\widehat{\gamma}_3 + \widehat{\gamma}_4 + \widehat{\gamma}_5) - 3(\widehat{\gamma}_1 + \widehat{\gamma}_2) \tag{4.21}$$

where $\widehat{\gamma}_k$ are obtained from the sample means as in Eq. (3.24). The mask arrangement is shown in Fig. 4.8(c). Once the shape function is determined, it is straightforward to determine whether an edge exists and, by extension, determine its location within the boundaries of the mask. Fig. 4.9(b) shows the result obtained by using a shape test in the location of the edge. It is interesting to compare this output to that of Fig. 4.7(b). The transition is now thinner. The shape test used in this section is by no means the only alternative to the usual F-test methodology. Kadar and Kurz[11] suggest an alternative approach, which they refer to as "post hoc comparison". It is based on Tukey's honestly significant difference test (HSD). In essence, it is a test between a pair of means based on multiple confidence intervals using the studentized range distribution. Since the Latin square model uses three treatments, the means that are considered here are the row, column, and diagonal means. Define $v = t_i - t_j$ and its estimate by $\widehat{v} = \widehat{t}_i - \widehat{t}_j$, then the confidence

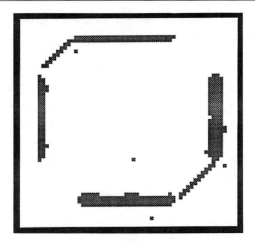

Figure 4.9. Output of Latin square coupled with shape test. (Shape test designed for $\theta = 45°$.)

interval for ν is

$$\widehat{\nu} - T_\nu \leq \nu \leq \widehat{\nu} + T_\nu \tag{4.22}$$

where $T_\nu = q_{\alpha,\nu} \sqrt{MS_e/m}$. Here, $q_{\alpha,\nu}$ are the percentage points of the studentized range with significance level α and $\nu = 5$ (the number of means under consideration). We assume that the size of the mask is $m = 5$.

4.5.2 Græco-Latin square-based detector

In the preceding section we introduced a detector structure capable of detecting edges in three directions. By using a mirror image of the diagonal mask, one can also detect edges in the 135° direction, but at the expense of an increase in the processing time. A more elegant approach is based on the GLS design, which was introduced in Section 3.6. We consider in this case four means, each representing a certain orientation, on which hypotheses can be based.

Recalling Eq. (2.68), the observations in the mask are parameterized by

$$\Omega : \begin{cases} y_{ij} = \mu + \alpha_i + \beta_j + \tau_k + \delta_l + e_{ij} & (i, j, k, l) \in S \\ \sum_{i=1}^m \alpha_i = 0; \sum_{j=1}^m \beta_j = 0; \sum_{k=1}^m \tau_k = 0; \sum_{l=1}^m \delta_l = 0. \\ (e_{ij}) \text{ independent } N(0, \sigma^2 \mathbf{I}) \end{cases} \tag{4.23}$$

The additional parameter δ_l models the diagonals in the 135° direction. A systematic 5×5 GLS design is shown in Fig. 4.10. The hypotheses representing the test of homogeneity of means are in this case

$$
\begin{aligned}
H_a &: \text{all } \alpha_i = 0 & \text{rows are homogeneous} \\
H_b &: \text{all } \beta_j = 0 & \text{columns are homogeneous} \\
H_c &: \text{all } \tau_k = 0 & \text{diagonals (45°) are homogeneous} \\
H_d &: \text{all } \delta_l = 0 & \text{diagonals (135°) are homogeneous}
\end{aligned}
\qquad (4.24)
$$

The corresponding F-tests are then given by Eqs. (2.76)–(2.79).

		Factor II			
Factor I	β_1	β_2	β_3	β_4	β_5
α_1	$A : \alpha$	$B : \beta$	$C : \gamma$	$D : \delta$	$E : \epsilon$
α_2	$E : \beta$	$A : \gamma$	$B : \delta$	$C : \epsilon$	$D : \alpha$
α_3	$D : \gamma$	$E : \delta$	$A : \epsilon$	$B : \alpha$	$C : \beta$
α_4	$C : \delta$	$D : \epsilon$	$E : \alpha$	$A : \beta$	$B : \gamma$
α_5	$B : \epsilon$	$C : \alpha$	$D : \beta$	$E : \gamma$	$A : \delta$

Figure 4.10. Systematic GLS design.

By introducing shape tests similar to Eqs. (4.18), (4.20), and (4.21), it is possible to locate the edge for which the hypothesis was rejected. Fig. 4.11 shows the result obtained by using the shape test.

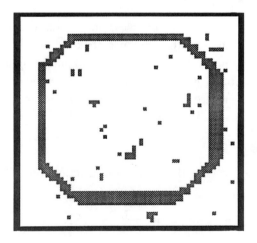

Figure 4.11. Output of Latin square coupled with shape test. (Shape test designed for $\theta = 45°$.)

4.6 Multidirectional detection in correlated noise

In this section we modify the Latin square-based detector so that it can operate well in both correlated and uncorrelated noisy environments. The motivation for designing such detectors is due to the presence of correlated noise structures in various applications where edge detection plays an important role. Among them is the feature detection problem, where detecting edges is the first step in the feature extraction process. Although similar detectors were introduced for line detection in Section 3.6 by using a GLS structure, the present section deals exclusively with the Latin square design. Needless to say, the same analysis can be carried out for the GLS design in edge detection. The assumption of a first-order Markov process for the noise process is not essential in the analysis, and a more general approach is possible without significant modifications.

The observations are represented by the following additive model

$$\Omega : \begin{cases} y_{ij} = \mu + \alpha_i + \beta_j + \tau_k + e_{ij} & (i, j, k) \in S \\ \sum_{i=1}^m \alpha_i = 0; \sum_{j=1}^m \beta_j = 0; \sum_{k=1}^m \tau_k = 0. \\ (e_{ij}) \text{ are } N(0, \sigma^2 \mathbf{K}_f) \end{cases} \qquad (4.25)$$

where $\mathbf{K}_f$ is assumed to be known or otherwise can be estimated from the available data.

Let us consider that $\mathbf{K}_f$ is separable into a product of row and column correlation matrices, respectively. Under this assumption $\mathbf{K}_f$ can be written in a tensor product form. By introducing a transformation matrix $\mathbf{P}$ such that $\mathbf{PP}^T = \mathbf{K}_f^{-1}$ (see Section 2.2.1), the sum of squares under the alternative becomes

$$SS_e(\mathbf{y}, \beta) = (\mathbf{y} - \mathbf{X}^T \beta)^T \mathbf{K}_f^{-1} (\mathbf{y} - \mathbf{X}^T \beta) \qquad (4.26)$$

To determine the sum of squares under the hypotheses, one needs to consider a similar form to that of Eq. (4.26) but with the design matrix $\mathbf{X}^T$ and the effect vector β modified to reflect the assumptions under the corresponding hypothesis. Stern and Kurz[10] derived the modified form of the Latin squares in the correlated noisy case. The reader is referred to reference [10] for a detailed discussion.

4.6.1 Sum of squares under Ω

The effect estimates are found by solving the usual normal equations. Solving $\partial SS_e(\mathbf{y}, \beta)/\partial \mu = 0$, we obtain

$$\sum_i \sum_j C_{i.} R_{.j} y_{ij} = C_{..} R_{..} \mu + R_{..} \sum_i C_{.i} \alpha_i + \sum_j R_{.j} \beta_j$$

$$+ \sum_i \sum_k C_{.i} R_{(k+1+m-i) \bmod(m), i} \tau_k \qquad (4.27)$$

Considering $\partial SS_e(\mathbf{y}, \boldsymbol{\beta})/\partial \alpha_i = 0$, we obtain

$$\sum_q \sum_j C_{iq} R_{.j} y_{qj} = R_{..} \sum_q C_{iq}(\alpha_q + \mu) + \sum_q \sum_j C_{iq} R_{.j} \beta_j$$

$$+ \sum_q \sum_k C_{iq} R_{.,(k+1+m-q) \bmod(m)} \tau_k$$

$$i = 1, 2, \ldots, m. \tag{4.28}$$

And by solving for the column effect the form $\partial SS_e(\mathbf{y}, \boldsymbol{\beta})/\partial \beta_j = 0$, we obtain

$$\sum_q \sum_i C_{.i} R_{jq} y_{iq} = C_{..} \sum_q R_{jq}(\beta_q + \mu) + \sum_q \sum_i C_{.i} R_{jq} \alpha_i$$

$$+ \sum_q \sum_k C_{.q} R_{j,(k+1+m-q) \bmod(m)} \tau_k$$

$$j = 1, 2, \ldots, m. \tag{4.29}$$

The solution of $\partial SS_e(\mathbf{y}, \boldsymbol{\beta})/\partial \beta_j = 0$ is somewhat more involved because we have to consider the arrangement of the mask under consideration. For the 45° diagonals of Fig. 4.5, we have the equation

$$\sum_q \sum_j \sum_i C_{qi} R_{(k+1+m-q) \bmod(m),j} y_{ij}$$

$$= \sum_i \sum_q \alpha_i C_{qi} R_{(k-1+m-q) \bmod(m),.} + \sum_i \sum_j \beta_j C_{i.} R_{(k+1+m-i) \bmod(m),j}$$

$$+ \sum_q \sum_i \sum_j C_{ij} R_{(k+1+m-i) \bmod(m),(q+1+m-j) \bmod(m)}(\tau_k + \mu)$$

$$k = 1, 2, \ldots, m \tag{4.30}$$

Eqs. (4.27)–(4.30) form a system of $3m + 1$ equations with $3m + 1$ unknowns. It can be shown by inspection that the system is singular. As a result, it is necessary to introduce side conditions on the parameters. One such choice is based on the following constraints

$$\sum_i C_{.i} \alpha_i = 0 \tag{4.31}$$

and

$$\sum_j R_{.j} \beta_j = 0 \tag{4.32}$$

Finally, the constraint on the diagonal effects is

$$\sum_i \sum_k \beta_j C_{i.} R_{(k+1+m-i) \bmod(m),.} \tau_k = 0 \tag{4.33}$$

Consequently, Eqs. (4.27)–(4.30) with the constraints Eqs. (4.31)–(4.33) can be written in the matrix form

$$
\begin{pmatrix}
\mathbf{A} & 0 & \mathbf{A}_c & \mathbf{M}_a \\
\mathbf{V}_a^T & 0^T & 0^T & 0 \\
0 & \mathbf{B} & \mathbf{B}_c & \mathbf{M}_b \\
0^T & \mathbf{V}_b^T & 0^T & 0 \\
\mathbf{C}_a & \mathbf{C}_b & \mathbf{C}_c & \mathbf{M}_c \\
0^T & 0^T & \mathbf{V}_c^T & 0 \\
0^T & 0^T & 0^T & m_\mu
\end{pmatrix}
\begin{pmatrix}
\alpha_1 \\ \vdots \\ \alpha_m \\ \beta_1 \\ \vdots \\ \beta_m \\ \tau_1 \\ \vdots \\ \tau_m \\ \mu
\end{pmatrix}
=
\begin{pmatrix}
\mathbf{A}_y \\ 0^T \\ \mathbf{B}_y \\ 0^T \\ \mathbf{C}_y \\ 0^T \\ \mu_y
\end{pmatrix}
\begin{pmatrix}
y_{11} \\ \vdots \\ y_{1m} \\ \vdots \\ y_{m1} \\ \vdots \\ y_{mm}
\end{pmatrix}
\tag{4.34}
$$

where the matrix components are given in the Appendix to Chapter 4.

In short matrix form Eq. (4.34) is expressed as

$$\mathbf{W}\beta = \mathbf{Zy} \tag{4.35}$$

and the estimate vector is then

$$\widehat{\beta} = \mathbf{W}^{-1}\mathbf{Zy} \tag{4.36}$$

Based on the assumption that the correlation matrices are known a priori, $\mathbf{W}^{-1}$ is computed before the scanning is started. Note that if the whole image does not have the same correlation matrix, it becomes necessary to recompute $\mathbf{W}^{-1}$ at each scan. The resulting sum of squares is found by substituting Eq. (4.36) in Eq. (4.26).

4.6.2 Sum of squares under the hypotheses

Under the hypotheses, the sum of squares is derived in a similar manner as before. To derive the effect estimates, the matrix components containing the effect on which the hypothesis is based are deleted from Eq. (4.34). As an example, we derive $SS_a(\mathbf{y}, \beta)$ under $H_a :$ all $\alpha_i = 0$. In matrix notation, the system is written as

$$
\begin{pmatrix}
\mathbf{A} & \mathbf{A}_c & \mathbf{M}_a \\
\mathbf{V}_a^T & 0^T & 0 \\
\mathbf{C}_a & \mathbf{C}_c & \mathbf{M}_c \\
0^T & \mathbf{V}_c^T & 0 \\
0^T & 0^T & m_\mu
\end{pmatrix}
\begin{pmatrix}
\beta_1 \\ \vdots \\ \beta_m \\ \tau_1 \\ \vdots \\ \tau_m \\ \mu
\end{pmatrix}
=
\begin{pmatrix}
\mathbf{A}_y \\ 0^T \\ \mathbf{C}_y \\ 0^T \\ \mu_y
\end{pmatrix}
\begin{pmatrix}
y_{11} \\ \vdots \\ y_{1m} \\ y_{m1} \\ \vdots \\ y_{mm}
\end{pmatrix}
\tag{4.37}
$$

In a matrix form, we have

$$\mathbf{W}_a \boldsymbol{\beta}_a = \mathbf{Z}_a \mathbf{y} \tag{4.38}$$

where $\mathbf{W}_a$ is a $(2m + 1) \times (2m + 1)$ matrix. The effect vector estimate under H_a is

$$\widehat{\boldsymbol{\beta}}_a = \mathbf{W}_a^{-1} \mathbf{Z}_a \mathbf{y} \tag{4.39}$$

The sum of squares is found by substitution. A similar technique is used to determine $SS_b(\mathbf{y}, \boldsymbol{\beta})$ and $SS_c(\mathbf{y}, \boldsymbol{\beta})$ under H_b and H_c, respectively.

Note that until now we did not use any particular form for the correlation matrix $\mathbf{K}_f$ except its tensored form. Stern and Kurz[10] discuss the implementation of the above structure for a Markov process with a tensored form, a structure that results in a considerable saving in processing time.

4.7 Edge reconstruction

In the course of detecting edges, certain situations occur such as those in severe noise environments, which result in the loss of edge information. One consequence of such a loss is difficulties in image segmentation problems where the edge contour is the basis for segmentation. Techniques for edge reconstruction are particularly useful in this case. More important are those reconstruction procedures capable of good performance in extremely noisy environments.

In the spirit of this chapter, we introduce one such technique based on the linear model. In edge reconstruction one must first specify the model for reconstruction. The reason is that many models can be specified depending on the class of noise distributions, degradation models, etc.

The present model was suggested by Behar.[14] He considers a partly degraded image that provides useful a priori knowledge. Certain regions of the image are corrupted substantially by additive noise with the possibility of additional structured background. The rest of the image is embedded in well-behaved noise. The well-behaved region of the image is used to estimate the parameters associated with the direction of the edges.

4.7.1 Methodology

The reconstruction process relies on concepts developed in previous sections. The maximum gradient edge detection concept used here was suggested by Aron,[22] Bose,[23] Kariolis,[24] and Eberlein.[15] The problem is described as follows. Consider the outline of an object whose edge is partially defined and is to be determined in full. Since we assume that a partial description of the edge is available, it makes sense to try to use this a priori knowledge to find the most probable direction and by extension the location of the next edge point. In a mathematical setting the problem is given the edge points set $E_0, E_1, E_2, \ldots, E_{n-1}, E_n$. We try to determine E_{n+1} such that C is satisfied. Here C defines a certain quality factor to be determined later. It is sufficient to say at this point that C is a quality criterion that might define the sharpness of the reconstructed edge, the thickness, etc.

A similar problem occurs in linear prediction, where in the presence of a finite set of data points or correlation points one wants to find the next point to minimize the mean square error. In the present case the quality criterion is difficult to formulate, given that we also have to take into account the visual appearance of the reconstructed edge.

The finding of the location of the next edge point E_{n+1} is accomplished by a maximum gradient search. The procedure is recursive in nature because edge points previously determined are used to reconstruct the next possible point.

Before we consider the reconstruction process, we review briefly the gradient method in image processing. A good discussion on this subject can be found in reference [14].

4.7.2 Gradient method

Consider that there are two coordinate systems (x, y) and (x_r, y_r) in the image plane. Here (x, y) and (x_r, y_r) correspond to the unrotated and rotated coordinates, respectively. The coordinate transformation is $x = x_r \cos \theta - y_r \sin \theta$ and $y = x_r \sin \theta + y_r \cos \theta$. The first partial derivative of the image function $f(x, y)$ in the (x_r, y_r) domain is then

$$\frac{\partial f(x, y)}{\partial x_r} = \frac{\partial f(x, y)}{\partial x} \cos \theta + \frac{\partial f(x, y)}{\partial y} \sin \theta$$

$$\frac{\partial f(x, y)}{\partial y_r} = -\frac{\partial f(x, y)}{\partial x} \sin \theta + \frac{\partial f(x, y)}{\partial y} \cos \theta \tag{4.40}$$

or in matrix form

$$\begin{pmatrix} \frac{\partial f(x,y)}{\partial x_r} \\ \frac{\partial f(x,y)}{\partial y_r} \end{pmatrix} = \begin{pmatrix} \cos \theta & \sin \theta \\ -\sin \theta & \cos \theta \end{pmatrix} \begin{pmatrix} \frac{\partial f(x,y)}{\partial x} \\ \frac{\partial f(x,y)}{\partial y} \end{pmatrix} \tag{4.41}$$

The direction in which $f(x, y)$ assumes maximum value is found by differentiation with respect to θ. From Eqs. (4.40), we obtain

$$\frac{\partial f(x, y)}{\partial x} \sin \theta + \frac{\partial f(x, y)}{\partial y} \cos \theta = 0 \tag{4.42}$$

and, solving for θ, we obtain the direction angle

$$\theta_0 = \tan^{-1} \left(\frac{\partial f(x, y)/\partial y}{\partial f(x, y)/\partial x} \right) \tag{4.43}$$

The maximum magnitude is

$$|G| = \sqrt{\left(\frac{\partial f(x, y)}{\partial x}\right)^2 + \left(\frac{\partial f(x, y)}{\partial y}\right)^2} \tag{4.44}$$

In the digital case, we consider finite differences and the digital gradient becomes

$$\sqrt{(\Delta_x f(x, y))^2 + (\Delta_y f(x, y))^2} \tag{4.45}$$

and

$$\tan^{-1}\left(\frac{\Delta_y f(x, y)}{\Delta_x f(x, y)}\right) \tag{4.46}$$

where

$$\Delta_x f(i, j) = f(i, j) - f(i - 1, j) \tag{4.47}$$

and

$$\Delta_y f(i, j) = f(i, j) - f(i, j - 1) \tag{4.48}$$

The principle underlying the use of the gradient in edge detection is the calculation of a derivative operator that discriminates between two regions of distinct gray levels. Therefore, using Eqs. (4.45)–(4.46) for every pixel in the image domain, it is possible to determine whether the pixel element belongs to the edge set. The edge orientation is determined by the angle in Eq. (4.46).

As is well known, the response of a difference operator is affected by the corrupting noise. As suggested in reference [14], one can smooth the picture before applying the operator or use a similar operator that is based on differences of local averages. In general, the gradient is expressed as the difference of means or, in a more general setting, as a weighted function of a subset of pixels. A drawback to this approach is that it is highly inefficient in an unspecified noise environment. Behar[14] suggests the use of directional masks, typified by the usual GLS structure, in addition to contrast techniques. A primary reason for this choice is the robustness of the F-test in the determination of the gradient.

Edge reconstruction
Because the main elements of the edge reconstruction procedure have been described elsewhere in this chapter, we limit ourselves to describing the edge searching technique.

Gradient search of edge
The search is initiated at a known edge point E_i as shown in Fig. 4.12.

Assuming that the edge direction is clockwise, the next possible edge element is on the right-hand planes of E_i, which are the points denoted by 1, 2, 3, 4, and 5.

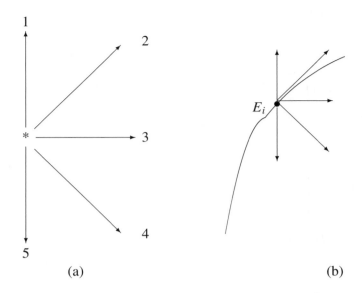

Figure 4.12. Gradient search of edge. (a) Localized search, (b) general clockwise
direction of edge.

This quantization of the edge points leads to the conclusion that the edge line can
move locally at directional increments of $45°$. The assumption is based on practical
consideration that the edge extends in the same direction for at least five pixels.
The gradient calculation is based on directional masks that are available at this
stage. Because we assume we have five directions, then for each pixel E_i, we have
gradients calculated from a 5×5 directional mask. The mask structure, shown in
Fig. 4.13, is directly related to the specific direction under consideration. Table 4.1
lists all possible directions. Masks D_1, D_3, and D_5 are squares of dimensions 5×5
while D_2 and D_4 are parallelepipeds in shape.

The data model in each case is the Græco-Latin square model. For each mask
we calculate two contrast functions of the form

$$\psi_1 = 2(\gamma_3 + \gamma_4 + \gamma_5) - 3(\gamma_1 + \gamma_2) \tag{4.49}$$

Table 4.1. *Mask directions.*

Point	Mask	Direction
1	D1	$+90°$
2	D2	$+45°$
3	D3	$0°$
4	D4	$-45°$
5	D5	$-90°$

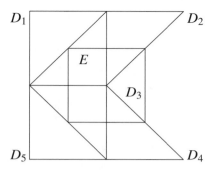

Figure 4.13. Masks templates for reconstruction.

and

$$\psi_2 = 2(\gamma_1 + \gamma_2 + \gamma_3) - 3(\gamma_4 + \gamma_5) \tag{4.50}$$

where γ determines a certain class of effects depending on $D_i, i = 1, 2, \ldots, 4$.

Since we obtain a contrast vector with ten components, the largest element will determine the location of the edge in addition to its direction. As in previous sections, the effects estimates are determined by the sample mean of the pixel elements along the appropriate effects under consideration. Figs. 4.14 and 4.15 show the various templates used in the contrast calculation phase.

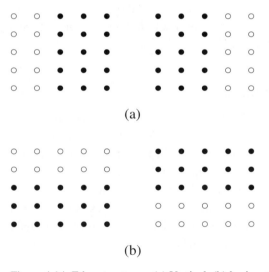

Figure 4.14. Edge structures. (a) Vertical, (b) horizontal.

A contrast comparison among all eight template contrasts is similar to a gradient comparison. The end result of this gradient-type search is the comparison of the maximum contrast to a fixed threshold. This will determine the direction of the

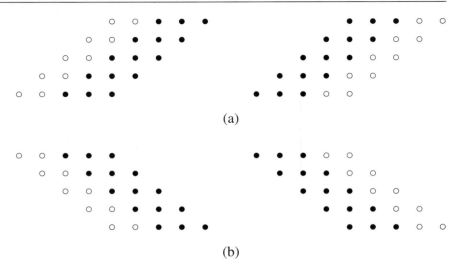

Figure 4.15. Edge structures. (a) Diagonal 45°, (b) diagonal 135°.

edge in a general confidence interval setting. From the theory of multicomparison techniques, this is analogous to comparing

$$|\psi_{max}| > (q\, F_{\alpha,q,n-r})^{1/2} s \qquad (4.51)$$

This also can be reformulated in an F-test of the hypothesis $H : \psi = 0$. If Eq. (4.51) is satisfied, then an edge is declared present and the pixel (i_0, j_0), within a given mask, is marked as an edge pixel. For a detailed discussion of this approach, the reader is referred to reference [14].

4.8 Concluding remarks

In this chapter several techniques for the detection of edges were introduced. Starting with the basic unidirectional design, which permits the detection of edges in vertical or horizontal directions, we proceeded to the more conventional bidirectional detector. The possibility of detecting two directions at the same time makes it an attractive alternative to the one-way design based detector.

A more sophisticated approach for multidirectional edges is introduced by considering the Latin and Græco-Latin designs. The inclusion of diagonals and the possibility of suppressing structured backgrounds are extremely important in complex scenes.

A modified version of the Latin square-based multidirectional detector is introduced to deal with the case of correlated noisy image scenes.

Edge reconstruction is introduced by using a gradient-based approach. Directional masks are used to formulate the gradient in the context of the GLS structure.

Appendix

By inspection, the various matrices in Section 4.6 are as follows.

$$\mathbf{A}(m-1\times m) = \begin{pmatrix} R_{..}C_{11} & \cdots & R_{..}C_{1m} \\ R_{..}C_{21} & \cdots & R_{..}C_{2m} \\ \vdots & \vdots & \vdots \\ R_{..}C_{m-1,1} & \cdots & R_{..}C_{m-1,m} \end{pmatrix} \tag{A.1}$$

$$\mathbf{A}_c(m-1\times m)$$
$$= \begin{pmatrix} \sum_q C_{1q} R_{.,(2-q+m)mod(m)} & \cdots & \sum_q C_{1q} R_{.,(2m+1-q)mod(m)} \\ \sum_q C_{2q} R_{.,(2-q+m)mod(m)} & \cdots & \sum_q C_{2q} R_{.,(2m+1-q)mod(m)} \\ \vdots & \vdots & \vdots \\ \sum_q C_{m-1,q} R_{.,(2-q+m)mod(m)} & \cdots & \sum_q C_{m-1,q} R_{.,(2m+1-q)mod(m)} \end{pmatrix} \tag{A.2}$$

$$\mathbf{B}(m-1\times m) = \begin{pmatrix} C_{..}R_{11} & \cdots & C_{..}R_{1m} \\ C_{..}R_{21} & \cdots & C_{..}R_{2m} \\ \vdots & \vdots & \vdots \\ C_{..}R_{m-1,1} & \cdots & C_{..}R_{m-1,m} \end{pmatrix} \tag{A.3}$$

$$\mathbf{B}_c(m-1\times m)$$
$$= \begin{pmatrix} \sum_q C_{.q} R_{1,(2-q+m)mod(m)} & \cdots & \sum_q C_{.q} R_{1,(2m+1-q)mod(m)} \\ \sum_q C_{.q} R_{2,(2-q+m)mod(m)} & \cdots & \sum_q C_{.q} R_{2,(2m+1-q)mod(m)} \\ \vdots & \vdots & \vdots \\ \sum_q C_{.,q} R_{m-1,(2-q+m)mod(m)} & \cdots & \sum_q C_{.,q} R_{m-1,(2m+1-q)mod(m)} \end{pmatrix} \tag{A.4}$$

$$\mathbf{M}_a(m-1\times 1) = \begin{pmatrix} R_{..}C_{1.} \\ R_{..}C_{2.} \\ \vdots \\ R_{..}C_{m-1,.} \end{pmatrix} \tag{A.5}$$

$$\mathbf{M}_b(m-1\times 1) = \begin{pmatrix} C_{..}R_{1.} \\ C_{..}R_{2.} \\ \vdots \\ C_{..}R_{m-1,.} \end{pmatrix} \tag{A.6}$$

$$
\mathbf{C}_a = \begin{pmatrix}
\sum_q C_{q1} R_{(2-q+m)mod(m),.} & \cdots & \sum_q C_{qm} R_{(2-q+m)mod(m),.} \\
\sum_q C_{q1} R_{(3-q+m)mod(m),.} & \cdots & \sum_q C_{qm} R_{(3-q+m)mod(m),.} \\
\vdots & \vdots & \vdots \\
\sum_q C_{q1} R_{(2m-q)mod(m),.} & \cdots & \sum_q C_{qm} R_{(2m-q)mod(m),.}
\end{pmatrix}
$$

(A.7)

Let $i' = (2 - i + m)mod(m)$ and $i'' = (2m - i)mod(m)$, then

$$
\mathbf{C}_b = \begin{pmatrix}
\sum_i C_{i.} R_{i',1} & \cdots & \sum_i C_{i.} R_{(i',m} \\
\sum_i C_{i.} R_{(3-i+m)mod(m),1} & \cdots & \sum_i C_{i.} R_{(3-i+m)mod(m),m} \\
\vdots & \vdots & \vdots \\
\sum_i C_{i.} R_{i'',1} & \cdots & \sum_i C_{i.} R_{i'',m}
\end{pmatrix}
$$

(A.8)

$$
\mathbf{C}_c = \begin{pmatrix}
\sum_i \sum_j C_{ij} R_{i',(2-j+m)mod(m)} & \cdots & \sum_i \sum_j C_{ij} R_{i',(2m+1-j)mod(m)} \\
\vdots & \vdots & \vdots \\
\sum_i \sum_j C_{ij} R_{i'',(2m-j)mod(m)} & \cdots & \sum_i \sum_j C_{ij} R_{i'',(2m+1-j)mod(m)}
\end{pmatrix}
$$

(A.9)

$$
\mathbf{V}_a^T = \begin{pmatrix} C_{.1} & C_{.2} & \cdots & C_{.m} \end{pmatrix}
$$

(A.10)

$$
\mathbf{V}_b^T = \begin{pmatrix} R_{.1} & R_{.2} & \cdots & R_{.m} \end{pmatrix}
$$

(A.11)

$$
\mathbf{V}_c^T = \begin{pmatrix} \sum_i C_{i.} R_{(2-i+m)mod(m),.} & \cdots & \sum_i C_{ij} R_{(2m+1-i)m/od(m),.} \end{pmatrix}
$$

(A.12)

$$
\mathbf{A}_y(m - 1 \times mn)
$$
$$
= \begin{pmatrix}
C_{11} R_{.1} & \cdots & C_{11} R_{.m}, & \cdots & C_{1m} R_{.1} & \cdots & C_{1m} R_{.m} \\
\vdots & & \vdots & & \vdots \\
C_{m-1,1} R_{.1} & \cdots & C_{m-1,1} R_{.m}, & \cdots & C_{m-1,m} R_{.1} & \cdots & C_{m-1,m} R_{.m}
\end{pmatrix}
$$

(A.13)

$$\mathbf{B}_y(m-1 \times mn) = \begin{pmatrix} R_{11}C_{.1} & \ldots & R_{1m}C_{.m} \\ \vdots & \vdots & \vdots \\ R_{m-1,1}C_{.1} & \ldots & R_{m-1,m}C_{.m} \end{pmatrix} \tag{A.14}$$

$\mathbf{C}_y(m-1 \times mn)$

$$= \begin{pmatrix} \sum_q C_{q1} R_{(2-q+m)mod(m),1} & \sum_q C_{q1} R_{(2-q+m)mod(m),2} & \cdots \sum_q C_{q1} R_{(2-q+m)mod(m),m} \\ \vdots & \vdots & \vdots \\ \sum_q C_{q1} R_{(2m-q)mod(m),1} & \sum_q C_{q1} R_{(2m-q)mod(m),2} & \cdots \sum_q C_{q1} R_{(2m-q)mod(m),m} \end{pmatrix}$$

$$\tag{A.15}$$

$$m_\mu = R_{..}C_{..} \tag{A.16}$$

5

Object detection

5.1 Introductory remarks

One of the most interesting problems in image processing and computer vision is the detection of specific patterns or objects. The dimensionality of the problem is related to the primary needs of the experiment. For example, the problem is referred to as two-dimensional object detection in satellite picture processing problems and related areas. In this case, what is available is merely a projection of the object on a two-dimensional plane. Three-dimensional object detection relates to the case where multiple projections of the object are available, and one has to make a decision regardless of the viewing position. Examples of the latter class of problems are active vision problems such as those encountered in robotics, where the robot hand is to be directed to specific locations depending on the presence of targeted objects that are actively sought by imaging sensors, such as on-board cameras in the case of a moving robot. The shape of the object is presumably known and stored in a reference database.

The primary intent of such procedures for the detection of two-dimensional objects that are embedded in a noisy environment is a satisfactory performance over a wide range of noise conditions, in contrast to traditional methods that rely on concepts such as the matched filter. Statistical methods have been applied with success in related problems in the past decade. Specifically, problems dealing with line detection and edge detection have been thoroughly investigated using statistical procedures based on experimental designs and the ANOVA.

A major undertaking in object detection is the detection of known objects regardless of position, scale, and rotation in the available image. Until recently, most of the procedures devised to solve this problem have been mainly concerned with noise-free data or, at most, with levels of noise that can render the effect of noise negligible. In that respect, one would pay a heavy price in the number of false alarms that will arise when the noise effect will no longer be negligible.

In this chapter the one- and two-way layouts are used as a basis for developing procedures for the detection of objects in images with correlated/uncorrelated background noise. The approach is similar to that of the feature matching technique in that features from the test object are matched to the reference object that is stored in the database. A mask that contains the object and sufficient background pixels is used as the template reference for the scanning procedure. Specific features are extracted in terms of test statistics based on ANOVA. The number of features necessary for the detection are kept at a minimum. This would result in a low complexity of implementation of the detectors and a much faster processing of the data.

Another problem considered in this chapter is a form invariant detection procedure that is implemented through the use of complex conformal mapping techniques in conjunction with ANOVA-based detection procedures.

In this chapter we present a coherent approach to the two-dimensional object detection problem. We start with problems related to detection in uncorrelated and correlated noise backgrounds with one main assumption, that the orientation of the object is assumed to be fixed and known a priori. Although this is a severe limitation, nonetheless this approach is useful for such tasks as picking parts from conveyor belts and testing of parts in industrial applications where the orientation is already fixed and known. We proceed by extending these techniques to the rotation invariant detection problem, thus eliminating the limitations encountered in the past. The chapter is organized as follows.

Section 5.2 deals with the detection of objects in an uncorrelated environment. The detection of designated targets is performed in two stages. First there is a learning stage where the original data are transformed by means of a one-to-one transformation to yield three arrays: a background array, a target array, and the combination of both, which is referred to as the standard array. Features in terms of F-statistics of the arrays are then extracted. The detection stage is then based on the comparison between the features corresponding to the test object and the features corresponding to the reference object. Next, we adapt the procedure to handle correlated noisy environments. The issue of reduction of the processing time and the apparent complexity of the resulting processor is addressed by means of a reduced algorithm that combines the useful characteristics of the contrast algorithm, introduced in line detection problems, with the F-statistics. The result yields a generalized procedure that can be used in uncorrelated as well as correlated noisy environments.

In Section 5.3 we address the two-dimensional object detection problem by introducing the concept of linear contrast techniques in noisy environments. The template representing the designated target is partitioned into two types of arrays: the target arrays and background arrays. Each array is assigned a contrast value based on the effects within the array. The overall contrast that is associated with the template is the feature to be used in the detection phase. From the theory of the multicomparison techniques, a threshold is defined and used in comparing the test contrast to the reference contrast. An extension of this procedure is then introduced in terms of a much larger contrast space, thereby providing more test features.

In Section 5.4 the rotation invariant object detection problem is addressed by combining a complex conformal mapping with one of the detection algorithms developed in previous sections; that is, the collected data are first transformed into the space defined in terms of the rotation angle and radial distance specific to the conformal mapping. The obvious advantage of this mapping is the reduction of a possible rotation angle between a test object and the reference to a translation in the transformed space. A feature matching approach that is based on the previous procedures is then used to detect the test object and indicate the possible rotation angle.

5.2 Detection methodology

Detecting objects in digital images is of prime importance in numerous fields, especially in medicine, machine vision, and military applications. Of foremost interest in any problems of object detection is the recognition of known objects that are assumed to be embedded in a noisy environment. The field has received significant attention, and various techniques are readily available to handle the detection problem. However, most techniques were developed for cases that involve either low noise levels or noise-free images. This is a handicap in itself due to the rather high number of false alarms that result from applying such techniques in any actual detection problem.

One major approach that has been successfully used in the past is based on statistical procedures. It involves the use of a feature matching procedure in conjunction with an ANOVA-based model for the observations. Before delving further into the details of the techniques, it is worth presenting other approaches that have been used.

First, we distinguish between what we will refer to as original and transformed space techniques. Namely, original space-based techniques refer to the use of the original data or observations without any transformation prior to the detection phase. The transformed space-based techniques involve techniques where the data is transformed according to some rules before the actual detection takes place. Each approach has its own merits and flaws, and it would be nearly impossible to rank them without considering specific tasks for which we may be able to draw meaningful conclusions.

Moment invariant techniques are of prime importance for the class of original space techniques. Used primarily in rotation invariant procedures, perhaps the moment invariant technique is the most popular one among pattern recognition scientists and engineers.

First introduced by Hu[25] in pattern recognition, moment invariants are based on the theory of algebraic forms.[26] They have been used in aircraft identification[27] and ship discrimination.[28] Recently, more work has been done in this area, mainly by introducing circular harmonic functions expansion of the image function[29] and complex moments.[30] Further studies[30]–[32] also included the impact of using a large set of moment invariants in the recognition process and especially the redundancy inherent in this type of problem. Alternative solutions to using only space domain moment invariants have been introduced in the form of Fourier domain-based moment invariants.[33] This approach follows from the fact that high-order space domain moment invariants are believed to be highly susceptible to noise, and thus the need to supplement low-order space-based moment invariants by low-order Fourier domain moment invariants. At this stage it is important to mention that most of the theory was developed for the special case of noise-free images. This constraint constitutes a major handicap in real world application where noise necessarily affects the behavior of moment invariants. Among the work devoted to the noise effect, we mention Abu Mostafa and Psaltis[30] and Teh and Chin,[32] who estimated the impact of noise on the behavior of moment invariants.

For the transformed data approach, we mention the main technique that is based on what is referred to as the topological mapping strategies. This technique involves the use of a transformation with certain interesting properties, for example, transforming a rotation in the original domain into a translation in the transformed domain, thereby greatly simplifying the detection process with respect to any rotation of the original object. Interesting results were obtained and presented in references [34] and [35] by using complex conformal mapping.

A radically different approach is that developed by Chang and Kurz.[38] An interesting aspect of the technique is the combination of essentially two properties. These are the transformation of the data so as to separate the pixels belonging to the target and background areas, respectively, and the use of an ANOVA-based model in characterizing the data, thereby providing means for integrating the noise effect in the detection process.

Mohwinkel and Kurz[9] proposed some procedures for the detection of basic patterns embedded in noisy images. Further results were obtained by Kariolis and Kurz[37] in the area of object detection. Chang and Kurz[21, 38] extended the ANOVA model to two specific pattern detection problems, namely the trajectory and object detection problems. The first is implemented by using designs SBIB and the latter by using a learning procedure in conjunction with a detection procedure based on the F-statistics under the two-way model. Kadar and Kurz[39] developed a procedure for the processing of three-dimensional images through the use of replicated models under ANOVA. Recent work includes the results obtained by Kang[66] in the area of radial object detection and Benteftifa and Kurz[40–43,98] in form invariant object detection and related problems.

5.3 Transformation-based object detector

In this section we describe a transformation-based detection procedure in which a match between a test object and a reference is established. In principle it is similar to the template matching technique.[17] However, instead of comparing the test and reference objects directly, we compare a set of features that are associated with each object of the reference set. The underlying advantage is the reduction in the data storage because very few features are needed.

5.3.1 Uncorrelated data case

The procedure is a two-stage process: a learning and a detection stage. The learning stage consists of the transformation of the array of observations into a space of two distinct arrays; namely, the background and target arrays denoted by $\mathbf{T}$ and $\mathbf{B}$, respectively. Assume that $Y(i, j)$ is transformed to yield the arrays $\mathbf{B}$ and $\mathbf{T}$, each array being homogeneous within itself. Following Chang and Kurz,[38] the array $\mathbf{S}$ is called a standard array if the following conditions are satisfied at a significance level α: (1) $\mathbf{S}$ is heterogeneous, that is, $FS_c > F_\alpha$ and $FS_r < F_\alpha$; (2) $\mathbf{B}$ is homogeneous, that is, $FB_c < F_\alpha$ and $FB_r < F_\alpha$; (3) $\mathbf{T}$ is homogeneous, that is, $FT_c < F_\alpha$ and $FT_r < F_\alpha$; and (4) Contrast exists between columns k and $k + 1$.

The tabulated thresholds F_α depend on the dimensions of $\mathbf{T}$, $\mathbf{B}$, and $\mathbf{S}$. As shown in Fig. 5.1, the learning process is the transformation of the template array into a global array $\mathbf{S}$ with size (p, q). The target and background arrays $\mathbf{T}$ and $\mathbf{B}$ have dimensions (p, l) and (p, q). Thus, to satisfy the above conditions, given a predetermined α, we seek the optimal values of p, q, and k. Hence, we can define the standard pattern by $\mathbf{S} = f(p, q, k, \alpha)$. The end result of the preceding characterization is to find a way to describe other standard arrays of which the three parameters p, q, and k are the same but the confidence levels are equal or higher than α. These patterns have the same visual effects, the difference being

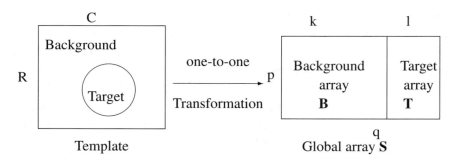

Figure 5.1. Transformation procedure.

in the relative intensity of the arrays **B** and **T**. During the detection process one compares the array **S** corresponding to test and reference templates. A meaningful comparison is obtained by considering the concept of visual equivalence. Define W as the transformation rules between original and transformed spaces. Then,

$$W : Y_a(m, n) > S_a(p, q)$$
$$W : Y_b(m, n) > S_b(p, q)$$

$$(5.1)$$

The arrays Y_a and Y_b are visually equivalent, with a confidence level α, if both S_a and S_b satisfy the conditions for standard patterns, that is, $S_a = (p, q, k, \alpha_a)$ and $S_b = (p, q, k, \alpha_b)$ with $\alpha_a \leq \alpha$ and $\alpha_b \leq \alpha$.

For each template of the database, one has to determine the optimum parameters of the standard array during the learning stage. The process is iterative in nature in that the process is stopped only when the conditions on F-statistics and contrast are satisfied. The primary step is the determination of transformation rules between original and transformed spaces, that is, mapping of pixel elements from template to standard array.

Let n_t and n_b be the total number of target and background pixels. Suppose that we can find a common nonrepeating factor d_i to n_b and n_t, then we can write

$$pq = d_i q = n_b + n_t$$

$$(5.2)$$

In which case, we have

$$p = d_i$$
$$q = (n_b + n_t)/d_i$$
$$k = n_b/d_i$$

$$(5.3)$$

Though in most cases it is not possible to find a common d_i, we can avoid this problem by starting with a minimal background, then increasing the number of background elements n_b until the conditions are satisfied. Note that by increasing the number of background elements, we also increase the contrast between target and background. Once the optimal (p, q, k) are found, the elements of the template are mapped to a position in the global array, that is, a location in **B** or **T**, depending on the nature of the element itself. The transformation rules along with the array dimensions are recorded as the features associated with the reference object.

During the detection stage the same technique is used. To compare a test template to the reference, it is necessary to make use of the concept of visual equivalence. That is, given (p, q, k, a) we can always have a standard pattern with (p, q, k, α_a), of which the confidence level α_a is less than or equal to α. Thus, by testing the calculated F-statistics and the contrast corresponding to the test template against the tabulated thresholds F_α, we can declare that the template is visually equivalent to the reference depending on whether the conditions are met. The detection stage is initiated using the following hypotheses

$$H : F_N = f(p, q, k, \alpha)$$
$$K : F_N \neq f(p, q, k, \alpha)$$

(5.4)

where α is the confidence level of the test, $f(p, q, k, \alpha)$ is the set of tabulated values of thresholds obtained during the learning stage, and F_N corresponds to the statistics of the test object. Fig. 5.2(b) shows the threshold values corresponding to the template in Fig. 5.2(a).

22	22	22	22	22	22	22	22	22	22
22	22	29	29	29	29	29	29	22	22
22	22	29	29	29	29	29	29	22	22
22	22	29	29	29	29	29	29	22	22
22	22	29	29	29	29	29	29	22	22
22	22	29	29	29	29	29	29	22	22
22	22	29	29	29	29	29	29	22	22
22	22	29	29	29	29	29	29	22	22
22	22	22	22	22	22	22	22	22	22

(a)

$$\text{TSr} = F(3,72) = 2.15 \qquad TSc = F(24,72) = 1.48$$
$$\text{TBr} = F(3,36) = 2.25 \qquad TBc = F(12,36) = 1.74$$
$$\text{TTr} = F(3,33) = 2.27 \qquad TTc = F(11,33) = 1.74$$

(b)

Figure 5.2. (a) Template in the learning stage, (b) threshold values, $p = 4$, $q = 254$, $k = 13$, and $\alpha = 10\%$.

5.3.2 Correlated data case

Until now, we considered the data to follow a normal distribution. When dependency exists among data points, the performance of the detector deteriorates, although not as seriously as in the case of the line and edge detectors. The main difference between the independent noise-based detector and the dependent noise-based detector is in the calculation of the F-statistics. Specifically, the sum of squares is now weighted by the respective correlation matrices. Since we assumed the two-way model in the parameterization of the data, the pixel elements in any of the three matrices of the transformation **B**, **T**, or **S**, respectively, are now described by

$$\Omega : \begin{cases} y_{ij} = \mu + \alpha_i + \beta_j + e_{ij} & i = 1, 2, \ldots, m; j = 1, 2, \ldots, n. \\ (e_{ij})N(0, \sigma^2 \mathbf{K}_T) \end{cases} \quad (5.5)$$

where α_i and β_j denote row and column effects, respectively. $\mathbf{K}_T$ is the correlation matrix of the noise components. Its exact form depends on the specific array of concern, namely, $\mathbf{S}$, $\mathbf{B}$, or $\mathbf{T}$. The noise components in the original space are dependent as given by the correlation matrix, which we denote by $\mathbf{K}_f$. We assume no particular form for $\mathbf{K}_f$ at this stage. Later on we specialize the results to the case of tensored form of $\mathbf{K}_f$. Note that the following analysis is valid for all three arrays.

5.3.3 Sum of squares under Ω

First, we consider the calculation of the sum of squares under the alternative Ω.

$$SS_e(\mathbf{y}, \beta) = \sum_{i=1}^{m} \sum_{j=1}^{n} \sum_{l=1}^{m} \sum_{k=1}^{n} (y_{ij} - \mu - \alpha_i - \beta_j) R_{ijlk}(y_{lk} - \mu - \alpha_l - \beta_k)$$

$$(5.6)$$

where $\mathbf{R}$ is such that $\mathbf{R} = \mathbf{K}_T^{-1}$. Solving the normal equations $\partial SS_e(\mathbf{y}, \beta)/\partial \mu = 0$; $\partial SS_e(\mathbf{y}, \beta)/\partial \alpha_a = 0$, $a = 1, 2, \ldots, m$ and $\partial SS_e(\mathbf{y}, \beta)/\partial \beta_b = 0$, $b = 1, 2, \ldots, n$, we obtain the system of equations

$$\widehat{\mu} = \frac{1}{R_{....}} \sum_{i=1}^{m} \sum_{j=1}^{n} R_{ij..} y_{ij}$$

$$\sum_{i=1}^{m} \sum_{j=1}^{n} R_{ija.} y_{ij} = \sum_{i=1}^{m} R_{i.a.}(\widehat{\mu} + \widehat{\alpha}_i) + \sum_{j=1}^{n} R_{.ja.} \widehat{\beta}_j \quad (5.7)$$

$$\sum_{i=1}^{m} \sum_{j=1}^{n} R_{ij.b.} y_{ij} = \sum_{j=1}^{n} R_{.j.b}(\widehat{\mu} + \widehat{\beta}_j) + \sum_{i=1}^{m} R_{i..b} \widehat{\alpha}_i$$

with the side constraints

$$\frac{1}{R_{....}} \sum_{j=1}^{n} R_{.j..} \widehat{\beta}_j = 0$$

$$(5.8)$$

$$\frac{1}{R_{....}} \sum_{i=1}^{m} R_{i...} \widehat{\alpha}_i = 0$$

In matrix form the above equations can be written as

$$
\begin{pmatrix}
\mathbf{A}_\alpha & \mathbf{A}_\beta & \mathbf{A}_\mu \\
\mathbf{B}_\alpha & \mathbf{B}_\beta & \mathbf{B}_\mu \\
\mathbf{V}_\alpha & \mathbf{0}^T & 0 \\
\mathbf{0}^T & \mathbf{V}_\beta & 0 \\
\mathbf{0}^T & \mathbf{0}^T & m_\mu
\end{pmatrix}
\begin{pmatrix}
\alpha_1 \\
\vdots \\
\alpha_m \\
\beta_1 \\
\vdots \\
\beta_n \\
\mu
\end{pmatrix}
=
\begin{pmatrix}
\mathbf{Y}_\alpha \\
\mathbf{Y}_\beta \\
0 \\
0 \\
\mathbf{Y}_\mu
\end{pmatrix}
\begin{pmatrix}
y_{11} \\
\vdots \\
y_{1n} \\
\vdots \\
y_{m1} \\
\vdots \\
y_{mn}
\end{pmatrix}
\tag{5.9}
$$

where the submatrices are given in the Appendix to Chapter 5.

Writing Eq. (5.9) in a compact form, we have $\mathbf{W}\beta = \mathbf{Z}\mathbf{y}$. With the side constraints shown in Eq. (5.8), the system of Eq. (5.7) is of full rank, which implies that $\mathbf{W}$ is invertible. Hence the vector of estimate $\widehat{\beta}$ is given by

$$
\widehat{\beta} = \mathbf{W}^{-1}\mathbf{Z}\mathbf{y} = \mathbf{A}_e\mathbf{y}
\tag{5.10}
$$

where $\mathbf{A}_e$ is equal to $\mathbf{W}^{-1}\mathbf{Z}$ and is defined as the effect matrix under the alternative. Since $\mathbf{K}_T$ is known a priori, the elements of $\mathbf{A}_e$ can be calculated beforehand and stored in a lookup table. Once the data vector $\mathbf{y}$ is collected, using Eq. (5.10) the estimate is obtained by applying $\mathbf{A}_e$ to $\mathbf{y}$. Thus, the total computation time is reduced considerably. The corresponding minimum of the sum of squares is then

$$
SS_e(\mathbf{y}, \beta) = (\mathbf{Q}_e\mathbf{y})^T \mathbf{K}_T^{-1}(\mathbf{Q}_e\mathbf{y})
\tag{5.11}
$$

where $\mathbf{Q}_e = \mathbf{I} - \mathbf{X}^T \mathbf{A}_e$. Here, $\mathbf{X}^T$ is the design matrix for the two-way model. The error matrix $\mathbf{Q}_e$ can also be computed before the actual processing takes place. The vector of estimates $\widehat{\beta}$ under each hypothesis is derived directly from Eqs. (5.10).

5.3.4 Sum of squares under $\omega_a = H_a \cap \Omega$

By eliminating the entries in Eq. (5.9) corresponding to the effects α_i, we obtain the matrix form

$$
\begin{pmatrix}
\mathbf{B}_\beta & \mathbf{B}_\mu \\
\mathbf{V}_\beta & 0 \\
\mathbf{0}^T & m_\mu
\end{pmatrix}
\begin{pmatrix}
\beta_1 \\
\vdots \\
\beta_n \\
\mu
\end{pmatrix}
=
\begin{pmatrix}
\mathbf{Y}_\beta \\
0 \\
\mathbf{Y}_\mu
\end{pmatrix}
\begin{pmatrix}
y_{11} \\
\vdots \\
y_{1n} \\
\vdots \\
y_{m1} \\
\vdots \\
y_{mn}
\end{pmatrix}
\tag{5.12}
$$

With $\mathbf{Q}_a = \mathbf{I} - \mathbf{X}_a^T \mathbf{A}_a$, we have

$$SS_a(\mathbf{y}, \beta) = (\mathbf{Q}_a\mathbf{y})^T \mathbf{K}_T^{-1}(\mathbf{Q}_a\mathbf{y}) \tag{5.13}$$

and the corresponding F-statistic is

$$F_r = \frac{(SS_a(\mathbf{y}, \beta) - SS_e(\mathbf{y}, \beta))/dfa}{SS_e(\mathbf{y}, \beta)/dfe} \tag{5.14}$$

where $dfa = m-1$ and $dfe = (n-1)(m-1)$ are the degrees of freedom associated with the quadratics in the numerator and denominator in Eq. (5.14), respectively.

5.3.5 Sum of squares under $\omega_b = H_b \cap \Omega$

Similarly, the deletion of the entries in Eq. (5.9) corresponding to the effects β_j yields the matrix form

$$\begin{pmatrix} \mathbf{A}_\alpha & \mathbf{A}_\mu \\ \mathbf{V}_\alpha & 0 \\ 0^T & m_\mu \end{pmatrix} \begin{pmatrix} \alpha_1 \\ \vdots \\ \alpha_m \\ \mu \end{pmatrix} = \begin{pmatrix} \mathbf{Y}_\alpha \\ 0 \\ \mathbf{Y}_\mu \end{pmatrix} \begin{pmatrix} y_{11} \\ \vdots \\ y_{1n} \\ \vdots \\ y_{m1} \\ \vdots \\ y_{mn} \end{pmatrix} \tag{5.15}$$

Letting $\mathbf{Q}_b = \mathbf{I} - \mathbf{X}_b^T \mathbf{A}_b$, we obtain the minimum sum of squares under H_b, which is given by

$$SS_b(\mathbf{y}, \beta) = (\mathbf{Q}_b\mathbf{y})^T \mathbf{K}_T^{-1}(\mathbf{Q}_b\mathbf{y}) \tag{5.16}$$

Finally, the corresponding F-statistic is

$$F_c = \frac{(SS_b(\mathbf{y}, \beta) - SS_e(\mathbf{y}, \beta))/dfb}{SS_e(\mathbf{y}, \beta)/dfe} \tag{5.17}$$

where $dfb = n - 1$ is the degree of freedom associated with the numerator.

It should be pointed out that $\mathbf{A}_a$, $\mathbf{A}_b$, $\mathbf{Q}_a$, and $\mathbf{Q}_b$ can also be calculated before the actual processing takes place, thus resulting in a considerable saving in processing time.

5.3.6 Determination of background and target correlation matrices

Consider the space of the original observations. We assume at this stage that the correlation matrix $\mathbf{K}_f$ is of tensored form. This permits further mathematical developments without undue complications. Therefore, $\mathbf{K}_f$ can be expressed in terms of a separable product of row and column correlation matrices. The general form is similar to that considered in Chapters 3 and 4, namely

$$\mathbf{K}_f = \mathbf{K}_r \otimes \mathbf{K}_c \qquad (5.18)$$

where $\mathbf{K}_r$ and $\mathbf{K}_c$ are given in Chapter 4.

Now, if we consider the space of the transformed observations, the correlation matrix, which we denote by $\mathbf{R}_S$, would be a randomized version of $\mathbf{K}_f$ due to the one-to-one transformation of the data. The evaluation of the statistics F_r and F_c using Eqs. (5.14) and (5.17) would prove to be both cumbersome and computationally difficult because R_s is no longer of a tensored form and, therefore, its inverse would be difficult to compute. However, due to the fact that the transformation is one-to-one, it is possible to derive $\mathbf{R}_S$ directly from $\mathbf{K}_f$.

Consider a vector representation of the observations obtained by a row stacking operation.[18] The corresponding vector is

$$\mathbf{Y}^T = [y_{11}, \ldots, y_{1c}, | \ \ldots \ |, y_{r1}, \ldots, y_{rc}] \qquad (5.19)$$

The transformation, which may be viewed as a relocation of the elements, will result in a similar vector $\mathbf{Y}_S^T$ but with the position information being drawn from the random transformation rules that were recorded during the learning phase.

Consider the permutation matrices $\mathbf{P}_r$ and $\mathbf{P}_c$, which assume the row and column positions, respectively, in the transformation. We can write

$$\mathbf{R}_S = \mathbf{P}_r \mathbf{K}_f \mathbf{P}_c \qquad (5.20)$$

Since $\mathbf{K}_f$ is symmetric, $\mathbf{P}_c$ is identically equal to $\mathbf{P}_r^T$, thus $\mathbf{R}_S = \mathbf{P}_r \mathbf{K}_f \mathbf{P}_r^T$.

Finally, it should be noted that the tensored structure of the matrix $\mathbf{K}_f$ is a somewhat restrictive assumption. In a more realistic model, the correlation between adjacent pixels tends to zero as the spatial separation increases.[18] As a result, most of the elements of $\mathbf{K}_f$ would be equal to zero. Therefore, techniques for the inversion of sparse matrices could be used in the determination of $\mathbf{K}_f^{-1}$.

The next step is the determination of the correlation matrices of the background and target arrays in the transformed space.

5.3.7 Background and target correlation matrices

The calculation of the sum of squares in Eqs. (5.13) and (5.16) under the hypotheses is conditional on the knowledge of the correlation matrices $\mathbf{R}_B$ and $\mathbf{R}_T$ of the background and target arrays, respectively.

Consider the sets ξ_B and ξ_T of background and target elements. Since both are subsets of the mixed observation set ξ_o, it is possible to infer the structure of both $\mathbf{R}_B$ and $\mathbf{R}_T$ directly from $\mathbf{R}_S$ using the same argument developed for the determination of $\mathbf{R}_S$ from $\mathbf{K}_f$. Thus, we can classify the elements of $\mathbf{R}_S$ in three categories: (1) correlation between elements of ξ_B, (2) correlation between elements of ξ_T, and (3) correlation between elements of ξ_B and ξ_T.

Using this information and the respective positions of the background and target elements, we can always transform $\mathbf{R}_S$ to obtain $\mathbf{T}$ so that

$$\mathbf{T} = \mathbf{U}_r \mathbf{R}_S \mathbf{U}_c = \begin{pmatrix} \mathbf{R}_B & \mathbf{H}^T \\ \mathbf{H} & \mathbf{R}_T \end{pmatrix} \tag{5.21}$$

where $\mathbf{U}_r$ and $\mathbf{U}_c$ are the required permutation matrices. The matrix $\mathbf{H}$ corresponds to the third type of correlation.

5.3.8 Structure of the permutation matrices

The permutation matrices are obtained from the elementary matrix $\mathbf{I}_N$ by permutation of the rows or columns depending on whether we consider row or column ordering of the data. We consider a column separation of the target and background arrays, so a row permutation is adopted. First, note that $\mathbf{U}_c = \mathbf{U}_r^T$, hence $\mathbf{U}_r$ can be obtained by permutation of the rows of $\mathbf{I}_N$ such that the top rows of U_r correspond to the position of the background pixels within the template. The lower part of $\mathbf{U}_r$ corresponds to the target pixels position. The following computer loop transforms the elementary matrix $\mathbf{I}_N$ to yield U_r[1]

for $i := 1$ **step** 1 **until** *total_pixels* **do**

    if $y_{ij} < threshold$

        $b\_counter := b\_counter + 1$

        $U(b\_counter) := i$

    else

        $t\_counter := t\_counter + 1$

        $U(offset + t\_counter) := i$

    endif

where *offset* is the total number of background pixels in the template. The permutation is stored in a vector form given that only one element of each row of $\mathbf{U}_r$ is nonzero. Fig. 5.3 shows several results under both scenarios, that is, the dependent and independent noise cases.

5.3.9 A reduced algorithm

The procedure developed in the previous section is essentially the test of a reference set as compared to the test values obtained at each scan of the image. For relatively

---

[1] We consider the vector $\mathbf{Y}_S^T$ for the determination of $\mathbf{U}_r$.

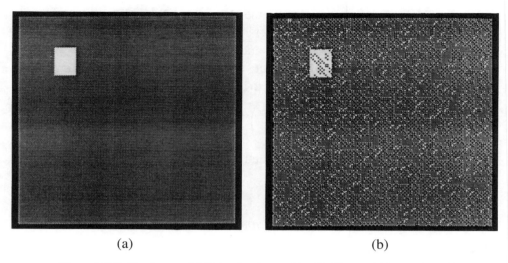

(a)   (b)

Figure 5.3(1). Test image. (a) Noise-free image 64×64, (b) $\sigma = 1.00$, $a = 0$, and $b = 0$.

small-sized objects, less than ten pixels wide, the extraction process results in a slight increase of computation time compared to the procedure based on the independent data case. For larger objects, the procedure becomes impractical, and an alternative in the sense of a reduced test set is sought.

During the learning phase the overall target-to-background pixel ratio is varied until the conditions are satisfied. However, note that we also test the contrast between column k and $k+1$ where column k is the separation between background and target arrays. This could be used as a basis for the detection procedure but with a proper modification to include all of the information contained in the arrays **B** and **T**. Therefore, instead of testing the contrast[2] between column k and $k+1$, we would test the overall contrast between **B** and **T**. However, we still must test for homogeneity and heterogeneity of **B**, **T**, and **S**, respectively, which can be accomplished using the test of the F-statistics corresponding to **S**.

For the array **S**, the contrast function is defined as the weighted sum of the effect estimates with the weights chosen so that their sum equals zero. Denoting the contrasts for the arrays **S**, **B**, and **T**, by ψ_S, ψ_B, ψ_T, respectively, we have $\widehat{\psi}_S = \widehat{\psi}_B - \widehat{\psi}_T$ where $\widehat{\psi}_B = \sum_{j=1}^{k} \widehat{\beta}_j^B$ and $\widehat{\psi}_T = \sum_{j=1}^{l} \widehat{\beta}_j^T$. The subscript here relates the effect to the array for which the contrast is calculated. Therefore, the overall contrast for the array **S** is $l\widehat{\psi}_B - k\widehat{\psi}_T$.

When used as a shape test, the contrast is tested whether it is equal to zero. When the contrast value exceeds zero, the object is detected. However, this threshold is

---

[2]Which should not be confused with the visual contrast.

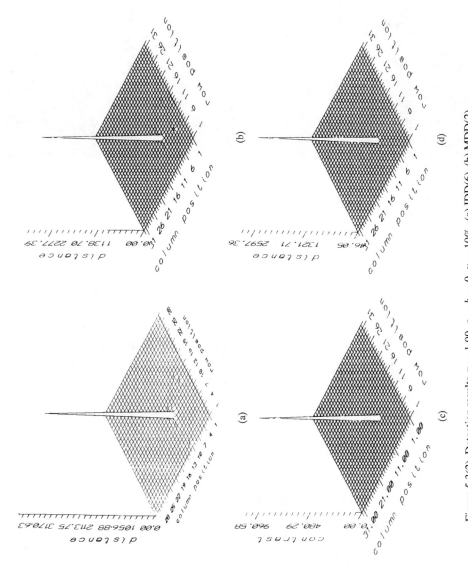

Figure 5.3(2). Detection results $\sigma = 1.00$, $a = b = 0$, $\alpha = 10\%$. (a) IDD(6), (b) MDD(2), (c) MDD(2 + C), (d) DDD.

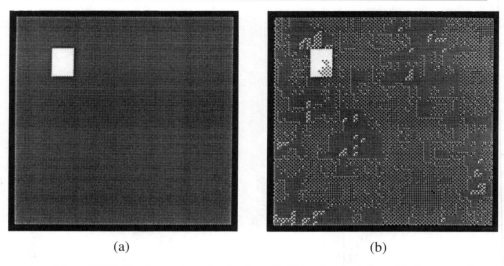

(a)   (b)

Figure 5.3(3). Test image (correlated noise). (a) Noise-free image 64×64, (b) $\sigma = 1.00$, $a = 0.85$, and $b = 0.85$.

not sufficient to reduce the false alarms resulting from closely related objects. Using Scheffe's technique for the S-contrast, the threshold for the one-dimensional case is defined as $T_H = (\widehat{\sigma}_\psi^2 F_{\alpha,1,n-r})^{1/2}$, where $\widehat{\sigma}_\psi^2$ is the estimate of the contrast variance and is given by $s^2 \mathbf{a}^T \mathbf{K}_T \mathbf{a}$. The estimate of the data variance s^2 is equal to $SS_e(\mathbf{y}, \beta)/(n - r)$, where $n - r$ is the rank of the quadratic form $SS_e(\mathbf{y}, \beta)$. The vector $\mathbf{a}$ is the vector of coefficients assigned to the effects in the contrast estimate. The contrast is then tested against a nominal contrast C using the interval $|\widehat{\psi}_S - C| \le T_H$.

The reduction in complexity of the present procedure is evident in the sense that only the contrast estimate needs to be calculated at each scan.

5.4 Partition-based object detector

The structure considered in the preceding section assumes two basic arrays: the target and background arrays. The separation does not take into account the specific properties of the object such as shape, internal holes, etc. In addition, it is implicitly assumed that the template has two dominant gray levels, which makes the separation into background and target pixels easy to obtain. For more complex objects, there is a need to increase the number of target and background arrays to accomodate the wide variations in pixel separation. In this section we present a technique that is based on the gray level properties of the object to be detected.

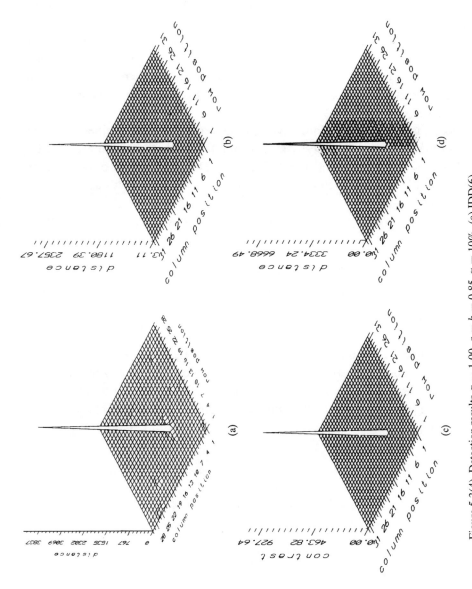

Figure 5.3(4). Detection results $\sigma = 1.00$, $a = b = 0.85$, $\alpha = 10\%$. (a) IDD(6), (b) MDD(2), (c) MDD(2 + C), (d) DDD.

5.4.1 Image representation

The image containing the target is assumed to be corrupted by zero mean Gaussian noise. This assumption is not restrictive, in the sense that the model accommodates deviations from the Gaussian distribution.

The collected data are represented by a two-dimensional rectangular array of pixels p_{lk}, where $l = 1, 2, \ldots, I$; $k = 1, 2, \ldots, J$. The gray level information at p_{lk} is denoted by y_{lk}. The image is scanned either sequentially or in parallel using a mask $\mathbf{Y}_{rc}$, where $r = 1, 2, \ldots, R$; $c = 1, 2, \ldots, C$. The conditions on the mask size are that $R \ll I$ and $C \ll J$.

Suppose the mask is partitioned, via a transformation to be defined later, into subarrays that if connected side-by-side form a layout that can be represented by $\mathbf{L}_{ij}$, where $i = 1, 2, \ldots, I_j$ and $j = 1, 2, \ldots, n$. The total number of observations in the mask is $\sum_{j=1}^{n} I_j$, which is also equal to $R.C$. Fig. 5.4 shows an example of such a layout.

The assumptions under the linear model, for a one-way design with fixed effects, are

$$
\Omega : \begin{cases} y_{ij} = \mu + \beta_j + e_i j & i = 1, 2, \ldots, I_j; j = 1, 2, \ldots, n \\ (e_{ij}) \text{ are independent } N(0, \sigma^2 \mathbf{I}) \\ \sum_{j=1}^{n} \beta_j = 0 \end{cases} \tag{5.22}
$$

where β_j denotes column effects. Each cell within the mask $\mathbf{L}_{ij}$ is the result of three terms: a common term to all cells that is denoted by μ and referred to as the general mean, and a vertical effect that is specifically related to observations aligned along columns in the layout. Finally, the term e_{ij} is included to account for the corrupting noise.

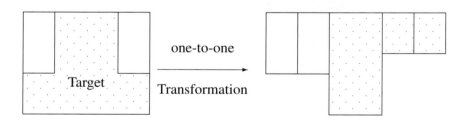

Figure 5.4. Template transformed to a one-way layout design.

The selection of the one-way design among all other possible designs, such as the two-way design or higher-order designs, is dictated by the need to have a model suited for the layout at hand, namely $\mathbf{L}_{ij}$. A two-way design, although very attractive in the sense that we could model vertical as well as horizontal orientations within the mask and at the same time retain a complexity of implementation that is similar to the one-way design, was ruled out because in general we cannot have I_j, the number of observations in each column of $\mathbf{L}_{ij}$, equal to a constant.

The hypothesis of main interest in this case is

$$H_o : \text{all } \beta_j = 0 \tag{5.23}$$

which means there are no effects under H_o.

The statistic for testing H_o is

$$F_a = \frac{\sum_{j=1}^{n} \sum_{i=1}^{I_j} (y_{..} - y_{.j})^2}{\sum_{j=1}^{n} \sum_{i=1}^{I_j} (y_{ij} - y_{.j})^2} \frac{(RC - n)}{n - 1} \tag{5.24}$$

Whenever F_a exceeds the threshold $F_{\alpha, n-1, RC-n}$, the hypothesis is rejected.

An unbiased estimate of the variance σ^2 in the model of Eq. (5.22) is the ratio $SS_e(\mathbf{y}, \beta)/m - r$, where $m - r$ is the rank of the quadratic form $SS_e(\mathbf{y}, \beta)$ and is identically equal to $RC - n$ under the assumptions of the one-way model. Here, $SS_e(\mathbf{y}, \beta)$ denotes the error sum of squares under the alternative and is equal to $SS_e(\mathbf{y}, \beta) = \sum_{j=1}^{n} \sum_{i=1}^{I_j} (y_{ij} - y_{.j})^2$.

When the hypothesis is rejected, it is necessary to include an additional test to locate the effect β_j that is not equal to zero. This is implemented using contrast functions.

Definition 5.1 *A parametric function ψ is defined as a linear combination of the effects $\beta_1, \beta_2, \ldots, \beta_n$ with known weights $(c_1, c_2, \ldots, c_n)$, that is,*

$$\psi = \sum_{j=1}^{n} c_j \beta_j \tag{5.25}$$

The choice of the coefficients is constrained by the location of the effect that is suspected of being significantly different from zero, that is, by associating an effect or a combination of effects with a physical characteristic of the pattern to be detected, we can determine whether this group of effects contrasts with the remaining effects in the layout. The condition on the parametric function is that it is estimable.

Definition 5.2 *The parametric function ψ is said to be an estimable function if it has an unbiased linear estimate, or*

$$E\left(\widehat{\psi}\right) = \psi \tag{5.26}$$

where the estimate is $\widehat{\psi} = \mathbf{c}^T \widehat{\beta}$.

Having defined the estimable function ψ and its unbiased estimate, the Gauss-Markov theorem establishes the construction and uniqueness of the estimate.

5.5 Basic regions and linear contrasts

The feature detection procedure is a two-stage process: a partition stage and a detection stage. During the detection stage, a contrast is associated with the feature based on a particular characteristic such as shape or gray level values.

The detection stage consists of the scanning of the image with a mask having the same dimensions as the template representing the feature. The feature calculated from the data collected at each scan is compared to the one derived during the partition stage. This forms the basis for the feature matching process.

At the heart of the technique itself is the transformation of the template into arrays of background and target arrays that are then arranged in a one-way layout. For bilevel images, two basic arrays are generally sufficient for the implementation of the contrast technique.[38, 41] However, in the case of graylevel images, we need to consider a much larger number of arrays to accomodate the variations in gray level in the template.

5.5.1 Histogram approach

This approach is based on the gray level information within the template. Fig. 5.5 shows the block diagram of the partition. First, a histogram layout of the template is performed to determine the dominant gray level peaks. In most cases at least two dominant peaks are found and indicate target and background regions. Using this information, we partition the template into homogeneous subregions.

The difference of intensity level between the target and background can then be defined in terms of the difference of contrasts of the regions of the partition.

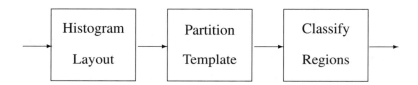

Figure 5.5. Histogram-based partition.

5.5.2 Linear contrasts

The contrast between the target and its surrounding background is related to the concept of linear contrast functions.

Definition 5.3 *A contrast among the effects $\beta_1, \beta_2, \ldots, \beta_n$ is a linear com-*

bination of all the $\beta'_j s$

$$\psi = \sum_{j=1}^{n} c_j \beta_j \qquad (5.27)$$

where the weights are subject to the condition $\sum_{j=1}^{n} c_j = 0$. The zero sum condition differentiates the Scheffe contrast technique from other related approaches such as the Bonferroni technique.[6]

Here ψ is an estimable function; therefore, its unbiased estimate is given by

$$\sigma_{\hat{\psi}}^2 = \sigma^2 \sum_{j=1}^{n} \frac{c_j^2}{I_j} \qquad (5.28)$$

where σ^2 is an unknown parameter whose estimate is given in Section 5.2. Note that the contrast variance is merely the variance of the data weighted by a factor that is dependent on the set of weights assigned to the effects.

The confidence interval of all values of the contrast ψ is defined as follows. First, we consider the special case where we have one contrast function ψ specified as a function of the effects $\beta_1, \beta_2, \ldots, \beta_n$. We use in this case a one-sided test of the contrast. The hypothesis considered is

$$H : \psi \geq C \qquad (5.29)$$

Let

$$t^* = \frac{\psi - C}{\sqrt{\sigma_{\hat{\psi}}^2}} \qquad (5.30)$$

Then, the hypothesis is accepted whenever t^* exceeeds the threshold $t(\alpha, n - r)$.

5.6 Detection procedure

The contrast technique in feature detection is implemented in a straightforward manner. For a given target, the contrast for a fixed background is compared to a nominal contrast C. If the test statistic value exceeds the threshold value, the target stands out and a hit occurs.

In the following section we develop a procedure under which we associate the contrast to a reference pattern. This is part of the learning stage and thus depends on the specific partition under consideration.

5.6.1 Contrast estimate

Definition 5.4 *Given a template represented by the array* $\mathbf{Y}_{rc}$, *where* $r = 1, 2, \ldots, R; c = 1, 2, \ldots, C$. *Under a suitable partition, there exists* M

background arrays denoted by $\mathbf{BC}_b$, $b = 1, 2, \ldots, M$ *and L target arrays* $\mathbf{TC}_t$, $t = 1, 2, \ldots, L$.

The procedure is first initiated by partitioning the template in a fixed number of regions that are then arranged to form a one-way layout. The contrast assignment is then accomplished through the following procedure.

Definition 5.5 *For a subarray of the partition, an elementary contrast is defined as the sum of the column effects. Thus, for a subarray with l column effects, the contrast is*

$$\xi = \sum_{j=1}^{l} \beta_j \tag{5.31}$$

and its estimate is given by

$$\widehat{\xi} = \sum_{j=1}^{l} \widehat{\beta}_j \tag{5.32}$$

Definition 5.6 *The total contrast ψ assigned to the template partitioned in M background arrays and L target arrays, with s_t and s_b being, respectively, the total number of target and background effects, is*

$$\psi = s_b \sum_{t=1}^{L} \xi_t^T - s_t \sum_{b=1}^{M} \xi_b^B \tag{5.33}$$

and its estimate is obtained by replacing the elementary contrasts by their estimates. Hence,

$$\psi = s_b \sum_{t=1}^{L} \widehat{\xi}_t^T - s_t \sum_{b=1}^{M} \widehat{\xi}_b^B \tag{5.34}$$

Note that the elementary contrast is not a contrast function, because the weight vector does not satisfy the zero-sum condition. A more rigorous definition of the elementary contrast would be a linear combination of the effects.

5.6.2 Threshold selection

The detection problem in the multidimensional contrast case is simply the test of the various hypotheses associated with the contrasts under consideration. Because we assume there is a wide variation in the gray level distribution among the arrays of the partition, we define two basic types of tests. First, we consider a hypothesis of the form $H_a : \psi_i \geq C_i$, which is used to determine the contrast between target and background arrays. Second, we consider $H_b : \psi_j \leq C_j$, which is used to test the homogeneity among the arrays of a certain class (either background or target arrays). It can be shown that increasing the number of contrasts increases the type II errors. Therefore, for the time being we consider the one-contrast case. An extension to the multicontrast case is deferred to the next section.

5.6.3 Contrast detection

The detection process is accomplished in the following steps. In step 1 the template at a given scan is partitioned into background and target arrays according to the partition rules obtained during the learning stage. In step 2 the column effects estimates for each array of the partition are calculated. The elementary contrasts are then determined. In step 3 the contrast estimate is computed as the sum of all elementary contrasts with the proper weighting factors so as to reflect the contribution in column estimates from each subarray. Then, the threshold is computed. The final step is the comparison of the test statistic value to the threshold for a confidence level α. If the contrast exceeds the prescribed threshold, the target is declared present. Otherwise the process is returned to step 1.

A ranking procedure is necessary to sort the most likely area containing the target. Noting that the sum of squares is minimum whenever the mask coincides with the object area, it is clear that the contrast variance is minimum whenever the object is detected. Therefore, the area associated with the smallest value of σ_ψ^2, and satisfying the detection conditions, contains the feature.

Summary of the contrast algorithm
A. Model
- $y_{ij} = \mu + \beta_j + e_{ij}$
- (e_{ij}) independent $N(0, \sigma^2)$

B. Definitions
- μ: general mean
- β_j: column effects
- (e_{ij}): noise component

C. Detection condition
- $t^* \geq t(\alpha, n - r)$

D. Algorithm
- D.1 Preprocessing
    Partition template in background and target arrays
    Determine contrast from elementary contrasts
    Determine nominal contrast C from training
- D.2 Detection
    Calculate threshold value
    Calculate contrast estimate based on column effects estimates
    Compare contrast estimate to threshold

Fig. 5.6 shows the result obtained by using the present procedure with two arrays – a background and target array – for both characters A and F. As expected, the patterns with the smallest threshold, that is, the minimum contrast variance, correspond to the desired objects to be detected. Fig. 5.7 shows the contrast variation for various values of the SNR. Here, the contrast remains within a fixed range for a wide range of the SNR values, which indicates the excellent characteristic of the contrast as a feature for detection.

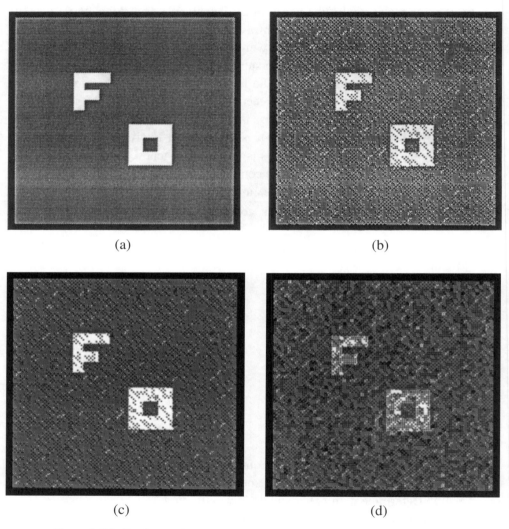

(a)   (b)

(c)   (d)

Figure 5.6(1). Test image 64×64. (a) Noise-free image, (b) noisy image $\sigma = 1.00$, (c) noisy image, $\sigma = 2.21$, (d) noisy image, $\sigma = 4.95$.

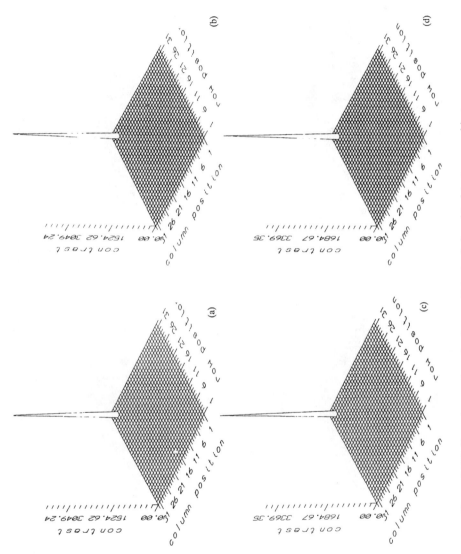

Figure 5.6(2). Detection results, mixed scene. Detector designed to detect square object. (a) $\sigma = 1.00$, $\alpha = 1\%$, (b) $\sigma = 1.00$, $\alpha = 10\%$, (c) $\sigma = 2.21$, $\alpha = 1\%$, (d) $\sigma = 2.21$, $\alpha = 10\%$.

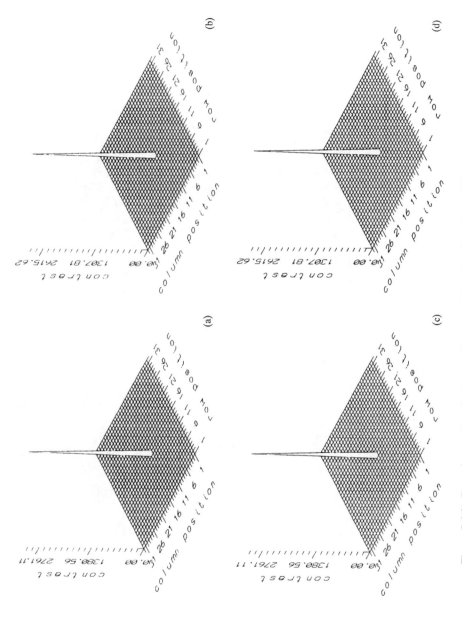

Figure 5.6(3). Detection results, mixed scene. Detector designed to detect letter "F."
(a) $\sigma = 1.00$, $\alpha = 1\%$, (b) $\sigma = 1.00$, $\alpha = 10\%$, (c) $\sigma = 2.21$, $\alpha = 1\%$, (d) $\sigma = 2.21$, $\alpha = 10\%$.

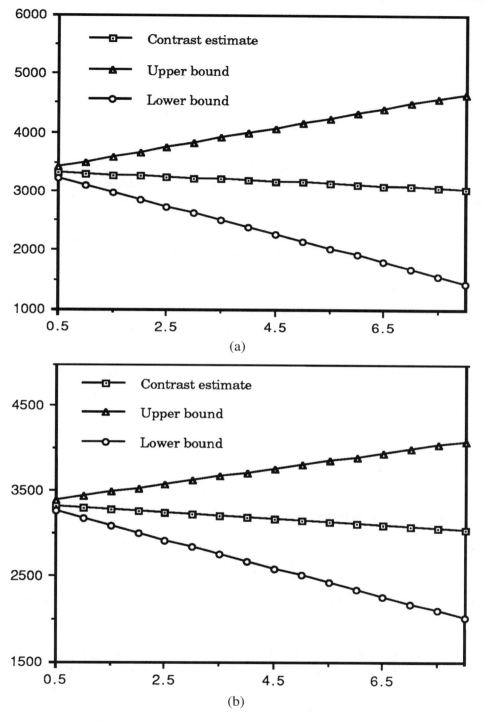

Figure 5.7. Contrast versus noise variance σ. Letter "F" (a) $\alpha = 1\%$, (b) $\alpha = 10\%$.

5.7 Orthogonal contrasts – an extension

The procedure developed in the preceding section uses only one feature in the detection process. In some cases it may be insufficient to use one contrast function in discriminating between closely related features that would have equal contrast values. In addition, although the contrast value is based on the specific partition, that is, feature template, one can always construct a completely different feature that will give rise to an identical contrast. A similar problem arises in image reconstruction, where given a finite number of moments,[44] one can always find an image that satisfies the constraints on the moments.

An interesting property of the contrast functions is that we can place an upper bound on the maximum number of *independent* contrasts that can be used in the description of the feature to be detected. This is to be contrasted with other techniques such as the one based on moment invariants, where it is often difficult to determine the optimum number of features needed in the detection process.

When using additional contrasts, the confidence level for each test is related to the family confidence level. For example, with α denoting the family confidence level, and assuming we have four tests based on the contrasts, the confidence level for each test is $\alpha/4$. Hence, we should expect an increase in the number of type II errors if we were to use additional contrasts.

> **Definition 5.7** *Two contrasts of means are said to be orthogonal if the sum of the product of their coefficients is equal to zero.*

In other words, given $\psi_1 = \sum_{j=1}^{p} c_{1j}\beta_{1j}$ and $\psi_2 = \sum_{j=1}^{p} c_{2j}\beta_{2j}$, ψ_1 and ψ_2 are orthogonal if the condition $\sum_{j=1}^{p} c_{1j}c_{2j} = 0$ holds.

> **Theorem 5.1** *In a one-way layout with p column effects $\beta_1, \beta_2, \ldots, \beta_p$, there are exactly $p - 1$ orthogonal contrasts.*

> *Proof:*   See Appendix to Chapter 5.

Because they are orthogonal, these contrasts are independent, that is, they contain all the information about the means and therefore the feature structure. Consequently, if we use all the contrast set in the feature detection procedure, all the information about the feature and its background in terms of contrast information is captured in the finite number of contrasts. In itself, this is a strong improvement vis-a-vis the single contrast approach.

An important remark at this stage is that the set of contrasts is not unique, although the orthogonal contrasts of each set are uniquely defined. The main point is that a set is generated by uniquely specifying the first contrast. All the remaining $p - 2$ contrasts are then successively generated from the previously calculated contrasts until the last one is found. Thus, if we were to select a different initial contrast, another set would be generated. However, in terms of contrast it will carry the same information as the preceding set.

> **Example** Consider $p = 4$. The two possible sets in terms of the weights are shown in Figs. 5.8 and 5.9. If we were to test β_3 against the remaining effects, that is, β_1, β_2, and, β_4, we would preferably select set C_1 because the first contrast already provides a way to test β_3 against the rest of the effects with equal weights assigned to them. The other contrasts in the set test the effects without taking into account β_3.
>
> In the general case of feature detection, we consider elementary contrasts rather than individual effects. With the total number of background and target arrays equal to $M + L$, the maximum number of independent contrast functions is $M + L - 1$. For example, for the array in Fig. 5.4, $l = 4$, $k = 2$, and $m = 2$, which yields the contrasts:
>
> $$\psi_1 = 4\xi_1^T + 4\xi_2^T + 4\xi_3^T - 8\xi_1^B - 8\xi_2^B$$
> $$\psi_2 = \xi_2^T - \xi_3^T$$
> $$\psi_3 = \xi_1^B - \xi_2^B \qquad (5.35)$$
> $$\psi_4 = 3\xi_1^T - 4\xi_2^T - 4\xi_3^T + \xi_1^B + \xi_2^B$$
>
> with $s_t = l + k + m = 8$ and $s_b = k + m = 4$. The contrast ψ_2 tests for the contrast between target arrays while ψ_3 tests for the contrast between background arrays.

Summary of the extended contrast algorithm

A. Model
- $y_{ij} = \mu + \beta_j + e_{ij}$
- (e_{ij}) independent $N(0, \sigma^2)$

B. Detection condition
- $\sum \psi_B \leq C_B$: Test of homogeneity among background arrays
- $\sum \psi_T \leq C_T$: Test of homogeneity among target arrays
- $\psi \geq C$: Overall contrast

C. Algorithm

	β_1	β_2	β_3	β_4
c_1	-1	-1	3	-1
c_2	1	-1	0	0
c_3	-1	-1	0	2

Figure 5.8. Contrast coefficients. Set A.

	β_1	β_2	β_3	β_4
c_1	-2	0	3	-1
c_2	0	-4	1	3
c_3	-13	7	-5	11

Figure 5.9. Contrast coefficients. Set B.

- C.1 Preprocessing
    Partition template into background and target arrays
    Determine vector of contrasts form from elementary contrasts
    Determine nominal contrast vector **C** from training
- C.2 Detection
    Calculate threshold values
    Calculate contrast estimates based on column effects estimates
    Compare contrast estimates to thresholds

5.8 Form-invariant object detector

In the previous sections we presented techniques for detecting objects in two-dimensional scenes using ANOVA techniques. The features used in the recognition process were based on F-statistics and contrast functions. The main objective of each procedure was to achieve reliable detection in the presence of noise. A major assumption in the detection process was that the test object must be properly aligned with respect to the reference object. In other words, the contrast calculation, in the case of the contrast-based algorithm, is performed according to the orientation of the object, creating a major problem when there is a rotation between test and reference objects.

A generalization of the contrast-based algorithm would of course include this rotation in the contrast calculation regardless of any rotation angle that might exist. In the literature, this is referred to as the 2-D rotation-invariant object recognition problem. Considerable attention has been devoted to it because of its numerous applications, especially in medical applications, machine vision, and target extraction.

Moment invariants have been used in solving this problem by many researchers.[25, 30, 31, 44] They relied on algebraic forms that, under certain combinations, are invariant to scale and rotation transformation of the data. A set of seven moments is usually derived for the test object and is compared to those corresponding to the reference object. Relevant results can be found in reference [44]. In this section we use a complex conformal mapping on the data before any procedure that we developed in previous sections is actually used. Complex conformal mapping has its foundations in the biological behavior of the human visual system.

Several studies have been conducted to understand the transformation occurring between the retinal plane and the cortical plane.

A viable mathematical model has been shown to be equivalent to a logarithmic mapping between the retinal plane of the eye and the cortex of the brain. Procedures based on this philosophy have been devised and used in the detection of two-dimensional objects irrespective of rotation and scaling of the data.[35, 45] This section makes use of the complex mapping in the early processing stage. As a result, the original data are transformed in such a way as to remove the effect of an existing rotation. We deliberately concentrate on the rotational invariance because we believe that it is more serious than the scaling effect. Needless to say, the scaling problem can be treated in the same manner. This section is organized as follows.

In the first subsection we review the complex conformal mapping as applied to invariant representation of the data. Next, an efficient detection technique is developed using one of the procedures developed in the previous sections in conjunction with the mapping of the data.

5.8.1 Invariant representation

Background

The existing procedures introduced in this chapter use a rectangular representation of the data points, that is, the location of each pixel is expressed in terms of the vertical and horizontal coordinates pair $(i, j) i = 1, ..., m; j = 1, ..., n$. This representation, that is, sampling, is inadequate if one is to take into account any possible rotation of the object. Therefore, an alternative representation in terms of polar coordinates is necessary for this case. Any point in the plane is then referenced by the pair (ρ, θ) where ρ is the radial distance from an appropriately chosen origin and θ is the polar angle. An invariant representation is then possible by using what is commonly referred to as complex conformal mapping. To understand the dynamics of the mapping, we first review some useful definitions.

Consider a point in the observation space that is referenced by

$$z = \rho e^{j\theta} \tag{5.36}$$

Now, in a logarithmic transformation of Eq. (5.36), we obtain

$$w = ln(z) = \ln \rho + j\theta \tag{5.37}$$

which can also be expressed as

$$w = \mathbf{u} + j\mathbf{v} \tag{5.38}$$

where $\mathbf{u} = \ln \rho$ and $\mathbf{v} = \theta$. In terms of the rectangular coordinates (i, j) the pair

$(\mathbf{u}, \mathbf{v})$ can be also expressed as

$$\begin{aligned} \mathbf{u} &= \ln(i^2 + j^2)^{1/2} \\ \mathbf{v} &= \tan^{-1}(j/i) \end{aligned} \tag{5.39}$$

Based on an analogy with the human visual system, the transformation is commonly referred to as the retino-cortical mapping between the retinal plane, represented in the present case by the polar plane (ρ, θ), and the cortical plane of the brain represented by the cartesian plane $(\mathbf{u}, \mathbf{v})$. Interesting properties that are associated with this particular mapping are easily obtained from Eqs. (5.36) and (5.37). In the following, the image in the polar plane is referred to as the retinal image and its transformed version as the cortical image.

Rotation invariance

Suppose that the retinal image represented by $\mathbf{Y}(\rho, \theta)$ is rotated by an angle θ_0. Thus, the corresponding pair $(\mathbf{u}', \mathbf{v}')$ is obtained by first considering

$$z' = \rho e^{j(\theta + \theta_0)} \tag{5.40}$$

and, using the logarithmic transformation, it implies

$$w' = \ln \rho + j(\theta + \theta_0) \tag{5.41}$$

and hence

$$\begin{aligned} \mathbf{u}' &= \ln \rho \\ \mathbf{v}' &= \theta + \theta_0 \end{aligned} \tag{5.42}$$

As a result, a rotation in the retinal plane results in a shift along the v-axis in the cortical plane.

Scale Invariance

Suppose the image $\mathbf{Y}(\rho, \theta)$ is scaled by a factor a. Then the corresponding point becomes

$$z'' = a\rho e^{j(\theta)} \tag{5.43}$$

which is transformed to the pair

$$\begin{aligned} \mathbf{u}'' &= \ln a\rho = \ln a + \ln \rho \\ \mathbf{v}'' &= \theta \end{aligned} \tag{5.44}$$

The scaling thus results in a shift by $\ln(a)$ along the u-axis. These simple properties have been used by Massone et al.[35] in the detection of objects for the case of binary images. An extension to the general case of gray scale images has been shown to be possible.[45]

The sampling of the retinal image is performed by using a nonuniform sampling grid that is formed by exponentially spaced concentric circles. Fig. 5.10 shows the mapping of concentric circles into vertical lines. The sampling is accompanied by a reduction in the image size. For a given partition of the original image, the number of concentric circles yields the maximum size of the u-axis, while the number of points on the circle determines the maximum size of the v-axis. The exact determination of (u_{max}, v_{max}) is based on such factors as the object size. Massone et al.[35] provide a detailed description of the procedure involved in the selection of the pair (u_{max}, v_{max}) for both the single and multiobject cases.

Detection procedure

The algorithm described here involves a two-stage procedure: a learning or training stage and a detection stage. In the training stage the useful features necessary for detection are extracted, while in the detection stage the comparison of the test features to the reference features is accomplished.

The first step during the learning stage is the mapping of the collected data onto the cortical plane. As a result, any existing rotation is reduced to a translation along the v-axis. The object to be detected is represented by the template $\mathbf{RT}(i, j)$ where $i = 1, ..., m; j = 1, ..., n$. The template obtained after sampling and mapping using Eqs. (5.36) and (5.37), denoted by $\mathbf{CT}(u, v)$, is then stored as the new object template.

Massone et al.,[35] in the binary image case, also computed the orientation of the principal axis of inertia as well as the position of the center of gravity in the cortical plane. The usefulness of these features stems from the fact that the origin of the v-axis is shifted to coincide with the orientation of the principal axis of inertia, and the center of gravity becomes the new origin. Their calculation in the noisy image case is, however, highly susceptible to the effect of the corrupting noise. In the present case the shift along the v-axis is detected by considering the projection of the cortical image on a cylinder. Hence, the v-axis can be thought of as being wrapped around the cylinder. Thus, a cyclic shift of the image will eventually recover the position of the image with respect to the shift.

Next, features needed to complete the detection are extracted. There are two possible procedures available to us: the contrast-based procedure of Section 5.6 and the transformation-based procedure of Section 5.5. The choice of either one is subjective, and the decision on which one to use is based on such factors as the size and shape of $\mathbf{CT}(u, v)$. Based on the contrast algorithm, we first partition $\mathbf{CT}(u, v)$ into background and target arrays. Then, we assign elementary contrasts to each following the technique outlined in Section 5.5. As a result, a contrast function can be calculated and is assigned to $\mathbf{CT}(u, v)$. On the other hand, if we use the transformation procedure of Section 5.5, the first step would involve the

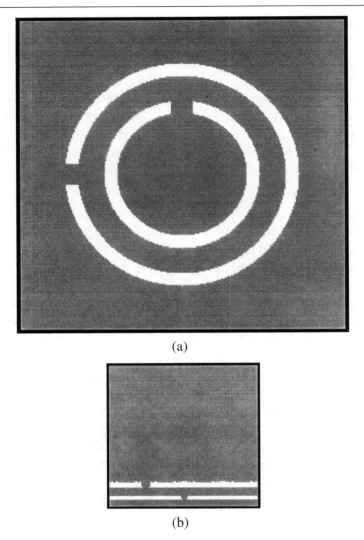

(a)

(b)

Figure 5.10. Conformal mapping of circles. (a) Original image in retinal plane, (b) image in cortical plane.

transformation of $\mathbf{CT}(u, v)$ to a space of background and target arrays. Features, in terms of row and column statistics, are then derived for a confidence level α. Fig. 5.11 shows the steps involved during the learning stage.

At this stage the information from the cortical image $\mathbf{CT}(u, v)$ has been transformed to a set of detection features characterizing the original object $\mathbf{RT}(i, j)$. Note that if the image contains multiple objects, the above learning stage needs to be repeated for each object.

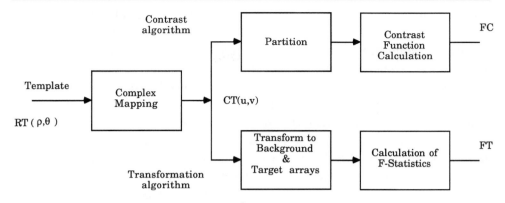

Figure 5.11. Learning stage.

Before we proceed with the description of the detection stage, we review the possible scanning schemes of the image with unknown objects. First, we consider a row scanning with a mask having the same size as the object template. Since each scan requires an additional cyclic scanning to detect an eventual shift, it would appear that the computation time will increase significantly. An alternate approach calls for an initial segmentation of the image to separate the various objects. The approach here is to segment the image first to isolate the various objects, then to perform a cyclic scanning to recover the rotation angle.

The detection phase is initiated with the following hypotheses

$$H : F = F_N$$
$$K : F \neq F_N$$

(5.45)

where F_N is the reference set of features that depends on the specific detection procedure.

Assume that the image has already been segmented, with the set of objects $O_1, O_2, \ldots, O_n$. Each object is then mapped using Eqs. (5.36) and (5.37). Using a mask with the same dimensions as the cortical reference object, a cyclic scanning is initiated for each object. At each pass, the feature set (that is, $FO_i, i = 1, 2, \ldots, n$) is extracted and compared to the reference set. If the set satisfies the detection condition, the object is detected and its position recorded. The rotation angle existing between the test and reference objects is also obtained after each successful detection.

5.9 Concluding remarks

In this chapter we first presented a transformation-based object detection procedure. We made use of the concept of visual equivalence within the framework of standard arrays. Next, the same procedure was adapted to the case of correlated data. The main problem in this case, namely the derivation of the correlation matrices corresponding to target, background, and global arrays, is easily solved by using permutation matrices to recover the transformation of element locations between original and transformed data spaces. The test statistics are then obtained by using the two-way layout parameterization in modeling the data.

A reduced form of the procedure is then introduced to deal specifically with the case of objects of large size. This is necessary because of the relatively long computation time required to find the inverses of the correlation matrices. A contrast function, defined in terms of the effects in the target and background arrays, is introduced as the additional test statistic in conjunction with the column and row statistics for the test of heterogeneity of the global array.

Then a contrast-based procedure is introduced. The object to be detected is partitioned into background and target arrays. Based on the contrast function, elementary contrasts are assigned to each array of the partition. Then a contrast function, defined as a linear combination of the elementary contrasts, is associated with the object. Next we provide an extension of the procedure by introducing the notion of orthogonal contrasts. As a result, more features are associated with the object, which can lead to more discrimination capability between closely related objects.

Finally, the rotation-invariant problem is addressed by introducing an invariant representation of the data. We use a complex conformal mapping of the data, thereby reducing rotation to translation, followed by one of the detection procedures introduced in previous sections. In addition to the detection of objects, we can detect the angle between the test and reference objects. This is a notable improvement over existing procedures that for the most part assume noise-free or very low noise levels. Typical results are given in Tables 5.1 and 5.2 in conjunction with Fig. 5.12.

Table 5.1. *Detection results. Letter "I."* $\alpha = 1\%$, $\sigma = 1.00$.

Rotation angle	Contrast	Threshold
0	4371.55	85.96
15	4480.49	86.16
45	4487.49	85.57
90	4407.64	85.79

Table 5.2. *Detection results. Letter "I."* $\alpha = 1\%$, $\sigma = 2.21$.

Rotation angle	Contrast	Threshold
0	4396.49	114.84
15	4505.43	114.46
45	4512.43	113.67
90	4432.58	114.27

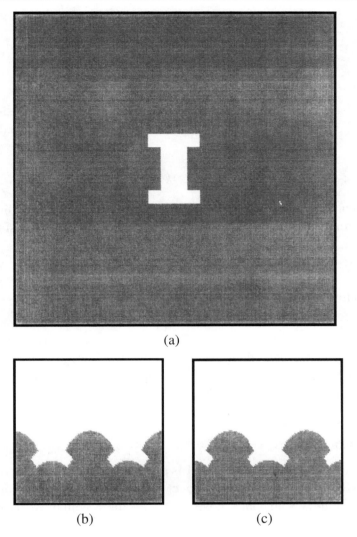

Figure 5.12. Conformal mapping of test image "I." (a) Original image in retinal plane, (b) image in cortical plane $\theta = 0$, (c) image in cortical plane $\theta = 90$.

Appendix

Contrast estimate calculation

We outline the general procedure for calculating the contrast estimate and its confidence interval given a particular partition. We assume here that the target and background arrays have already been obtained. The template under consideration is shown in Fig. 5.4. The one-way layout is obtained by placing the arrays side by side.

A.1 Contrast calculation

There are $k + m$ background effects and $l + k + m$ target effects. Thus, by using the elementary contrasts $\xi_1^T, \xi_2^T, \xi_3^T$, and ξ_1^B, ξ_2^B, the final expression for the contrast is given by

$$\psi = (k + m) \left(\xi_1^T + \xi_2^T + \xi_3^T \right) - (l + k + m) \left(\xi_1^B + \xi_2^B \right) \tag{A.1}$$

where the elementary contrasts are calculated following the technique outlined in Chapter 5. The contrast estimate is then obtained by replacing each contrast in Eq. (A.1) by the corresponding estimate.

A.2 Contrast variance

An essential component in the contrast variance calculation is the data variance estimate. It is identically equal to $SS_e(\mathbf{y}, \beta)/m - r$ where $m - r$ is the rank of the quadratic form in the numerator. Since the one-way design is used in the parameterization, it can be shown that for the template in Fig. 5.4 $m - r$ is equal to $(p + q)(k + m + l) - (2k + 2m + l + 1)$.

Orthogonal contrasts

Here the outline of the proof of Theorem 5.1 is presented.

Suppose we have selected the weight vector $\mathbf{c}_1$ corresponding to the first contrast. Then, to determine the second contrast, we must solve the system of equations

$$\sum_{j=1}^{p} c_{2j} = 0 \tag{A.2}$$

and

$$\sum_{j=1}^{p} c_{2j} c_{1j} = 0 \tag{A.3}$$

Eq. (A.2) is the zero-sum condition while (A.3) is the orthogonality condition.

To prove there are only $p - 1$ orthogonal contrasts, it is sufficient to prove that the weight vector associated with the pth contrast is identically equal to the zero vector.

To that end, consider the system of equations obtained from applying the orthogonal and zero-sum conditions up to the pth order. We obtain,

$$\sum_{j=1}^{p} c_{pj} c_{1j} = 0$$
$$\sum_{j=1}^{p} c_{pj} c_{2j} = 0$$
$$\vdots \qquad\qquad\qquad\qquad (A.4)$$
$$\sum_{j=1}^{p} c_{pj} c_{(p-1)j} = 0$$
$$\sum_{j=1}^{p} c_{pj}$$

which in matrix form can be written as

$$\mathbf{A}\mathbf{c}_p = \begin{pmatrix} c_{11} & c_{12} & \cdots & c_{1p} \\ c_{21} & c_{22} & \cdots & c_{2p} \\ \vdots & \vdots & \vdots & \vdots \\ 1 & 1 & 1 & 1 \end{pmatrix} \begin{pmatrix} c_{p1} \\ c_{p2} \\ \vdots \\ c_{pp} \end{pmatrix} = \begin{pmatrix} 0 \\ 0 \\ \vdots \\ 0 \end{pmatrix} = 0 \qquad (A.5)$$

The rows of $\mathbf{A}$ are linearly independent because of the orthogonality condition. Consequently, rank($\mathbf{A}$) = p, which implies that Eq. (A.5) holds only when $\mathbf{c}_p = (c_{p1}, c_{p2}, \ldots, c_{pp}) = \mathbf{0}$. QED.

6

Image segmentation

6.1 Introduction

In some applications such as feature detection, the initial step before the detection is the segmentation of the image into various regions to separate the feature from the background. This procedure is commonly referred to as image segmentation. Depending on whether there are single or multiple features, the result is a partition of the image into a certain number of homogeneous regions. Each pixel element of the image is assigned to one of the homogeneous regions. Some criteria of region homogeneity are usually gray level intensity, color, texture, etc. Hence, image segmentation can be regarded as scene classification with respect to some criteria. The process is complicated most of the time by essentially two problems: the nonuniformity of the gray level intensity of the image feature regions and the loss of contrast in some of the regions.

A popular approach to segmentation is based on region growing, which involves the merging of small uniform regions to form large regions without the uniformity of the combined regions being violated. The result of the merging process in this case depends on a suitable uniformity criterion. Some techniques in this area are based on estimation theory.[63] The region-based segmentation procedures are classified into three basic categories: pure splitting, pure merging, and split-and-merge. A survey of these three schemes can be found in reference [46]. In addition, it is recommended to consult references [47]–[68].

Segmentation strategies using a region-growing approach are hierarchical in nature. The first step in the pure merging divides the image into small regions, which are merged in the second step following a particular uniformity criterion. In the pure split procedure, the image is split into successive regions if the region under consideration is not uniform. Finally, the split-and-merge procedure is a two-step process that first divides the image into many subregions. A test of homogeneity is carried out at each level of the partition. Whenever the test is rejected, the subregion is further divided. The second step is the merging of the segments that

128

are adjacent and satisfy the uniformity criterion. In recent years the split-and-merge procedure has proven to be the method of choice in most segmentation problems. As mentioned before, the region uniformity test plays an important role in determining the stopping point in the partition of any given region. Therefore, most of the work in recent years has revolved around the development of adequate homogeneity test criteria.[46] Three problems hinder the development of a multipurpose segmentation algorithm: (1) the loss of contrast in some of the regions in the image, (2) region size, which sometimes results in the loss of the confidence level in statistical-based tests, and (3) corrupting noise factor. While the first two problems have been well studied over the past two decades,[63, 69] the effects of the corrupting noise have at best been marginalized. Even though in some areas the assumption of a noise-free image might be true, for most other applications the noise term has an important corrupting effect such as feature masking. This necessitates a robust segmentation procedure that includes the noise term explicitly in the data model and, by extension, the segmentation itself. The data model used in this chapter, namely the nested design model, includes the noise term in the parameterization of the observations. We assume in the present case i.i.d observations. The dependent data case is more involved and requires some complex matrix manipulation that is left for future research.

The chapter is organized as follows. We first discuss in detail the split-and-merge segmentation procedure. Next, we present the nested design model on which the region-growing algorithm is based. In particular, we discuss the fixed effects nested design model. The region testing and class formation are presented in Sections 6.4 and 6.5.

Nomenclature

The following are definitions of terms used throughout the chapter.

Tile: An image is divided into overlapping square tiles. A tile constitutes the first level of the partition.

Region: A tile, when it is found to be nonuniform, is divided into four equal square regions.

Quadrant: The basic constituent of a region. Each region, when it is found to be nonuniform, is divided into four equal square quadrants.

Segment: A collection of quadrants is a segment.

6.2 Segmentation strategy

Region growing is the formation of large meaningful regions from a large number of small regions. Assume that the image is of size $I \times J$. Each pixel in the image domain is referenced by the coordinates (i, j) where $i = 1, 2, \ldots, I$ and $j = 1, 2, \ldots, J$. The gray level information at location (i, j) is denoted by y_{ij}. A localized approach to the segmentation is considered here, where instead of using the entire image as the initial partition, overlapping square regions of size $M \times M$ are chosen for the initial partition. This is similar to the tile-by-tile approach described in reference [48], although the reason behind the choice of the approach in the present context has more to do with the region size, that is, the sample size in the statistical description of the data, rather than the hardware limitation of reference [48]. The localized nature of ANOVA, where masks of small sizes are used in the processing, has been described already in reference [9]. For the simulations described in this chapter, the initial cut size M for the square regions has been chosen equal to 8 with the smallest region size in the splitting stage equal to 2. It is understood that other values can be chosen, although the value $M = 16$ appears to be, after simulation, the maximum value that allows meaningful results. A tile in the present segmentation strategy consists of four regions, each of size $(M/2) \times (M/2)$. Each region is also formed of four quadrants, each of size $(M/4) \times (M/4)$.

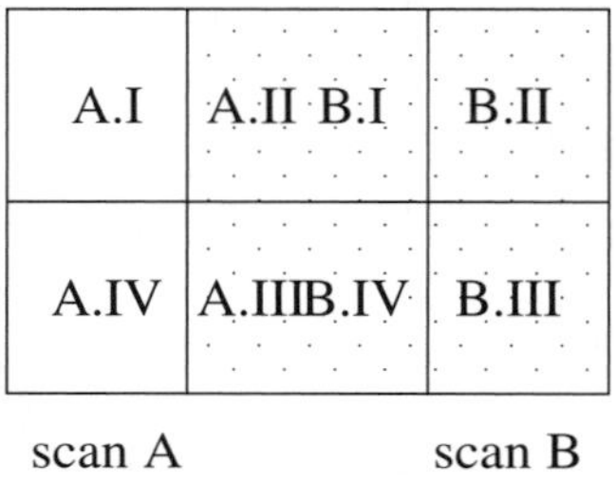

A.I	A.II B.I	B.II
A.IV	A.IIIB.IV	B.III

scan A   scan B

Figure 6.1. Overlapping tiles in a row scanning. The junction is represented by regions {A.II and A.III} in scan A and {B.I and B.IV} in scan B, respectively.

Fig. 6.1 shows the concept behind the split-and-merge procedure. The procedure is described for a single tile. A test is carried out to determine whether the region is homogeneous. Here as well as at any level of the partition, the homogeneity test is carried out using column and row F-statistics. The same technique is described in reference [41] in the case of object detection. If the tile is found to be heterogeneous, it is split into its four equal regions. A tree with the second level consisting of four nodes is therefore generated. To determine which region has resulted in the rejection of the homogeneity test of the tile, additional F-statistics are calculated for each region. Instead of splitting the regions that are determined to be heterogeneous, a test is carried out to determine the primitive quadrants that are not compatible with the remaining quadrants of the heterogeneous region. Fig. 6.2 shows the concept behind the region testing. Note that we consider only adjacent segments in a binary procedure, that is, only two regions are tested at a time. The homogeneity criterion used here is again based on column and row F-statistics.

The second stage of the procedure involves the merging of the quadrants from each region with quadrants from neighboring regions. In this case, an adaptive

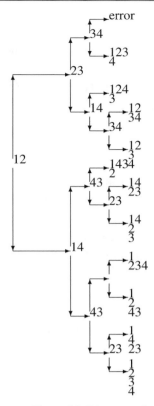

Figure 6.2. Binary quadrant testing.

nested design model is used for the data representation. The initial partition consists
of segments that are the grouping of uniform quadrants from the four regions
constituting the tile. Using a segment as the initial class in the segmentation,
segments from the partition are successively added to the design until all regions
are processed. At each iteration, two tests are carried out. The first is the region
test, which determines whether quadrants in a class are homogeneous. This is the
column test in the nested design parameterization. The second test is the row test
and determines whether the classes in the segmentation are disjoints.

The split-and-merge segmentation procedure proposed in this chapter is implemented in the following phases. Note that each class in the partition is implemented
in terms of linked lists. Several logical predicates are used in determining the homogeneity of the various elements in the segmentation.

Phase 1: Splitting the tile
Based on the predicate P_T, it is determined whether the tile is homogeneous. A split
in four regions is made if the predicate is false. Otherwise, the tile is considered as
a new class in the partition and therefore results in a new additional linked list.

Phase 2: Splitting the regions

Whenever the tile predicate is false, the tile is split into four regions R_i, $i = 1, 2, 3, 4$ of equal size $(M/2) \times (M/2)$. A region predicate P_R is invoked to determine the homogeneity of each region. If the predicate is false, the region of interest is further split in four quadrants Q_l, $l = 1, 2, 3, 4$, each of size $(M/4) \times (M/4)$. Otherwise, the region is declared homogeneous and results in a single linked list in the tile partition.

Phase 3: Merging the homogeneous quadrants of regions

For each heterogeneous region, the quadrants that are not uniform with the other quadrants are located. Toward this end, two quadrant predicates can be used and are defined in Section 6.5. Whenever the predicate is true, adjacent quadrants are merged. At the end of this phase, a number of linked lists are generated and correspond to the number of similar quadrants. Note that in terms of the number of linked lists, the minimum is 2 and the maximum is 4.

Phase 4: Merging adjacent quadrants from the different regions of the tile

The final processing of the tile is the merging of adjacent quadrants from the four regions in the tile. The predicate is defined in terms of F-statistics based on the nested design model. This is an iterative process started by adding one linked list, that is, a segment at a time to the initial reference segment.

Phase 5: Grouping adjacent segments from neighboring tiles

Because the present procedure uses a localized approach to the segmentation, it is necessary to group linked lists from adjacent tiles. This is facilitated by using overlapping tiles along each row and column. Therefore, in a row merging, the scanning mask is moved every $M/2$ columns to retain the second half of the previous tile as the first half of the new tile. Consequently, the partition found after processing the new tile is merged with the previous tile by locating the quadrants common to both. Fig. 6.1 shows the merging operation for two overlapping tiles. The same strategy is used in column merging.

6.3 Nested design model

To understand the mechanisms behind the adaptive merging in the final stage of the tile segmentation, we first review the nested design under ANOVA and its applications. We consider a fixed-effects model that is useful in gray level images. In the remaining part of the analysis, we assume a two-factor nested design with the factors designated by α and β.

6.3.1 Gray level images

Let y_{ijk} denote the kth observation when factor α is at the ith level and factor β is at the jth level. For each factor combination, it is assumed there are K_{ij} observations. The nested design when both α and β are fixed is the parameterization of the observations by

$$\Omega : \begin{cases} y_{ijk} = \mu + \alpha_i + \beta_{j(i)} + e_{ijk} \\ e_{ijk} \text{ independent} N(0, \sigma^2 \mathbf{I}) \\ i = 1, 2, \ldots, m; j = 1, 2, \ldots, J_i; k = 1, 2, \ldots, K_{ij} \end{cases} \tag{6.1}$$

where μ is the general mean, α_i and $\beta_{j(i)}$ are effects subject to $\sum_i \alpha_i = 0$ and $\sum_j \beta_{j(i)} = 0$ for all i, respectively. The error terms e_{ijk} are independent normal with zero mean and variance σ^2. The notation used in Eq. (6.1) follows the convention of the nested design.[2] Therefore, $\beta_{j(i)}$ denotes the specific level of β when α is at the ith level. Also, e_{ijk} denotes the kth error term for the (i, j) factor combination.

The motivation for using the present nested design model is that in some cases, some of the levels of β do not occur with some levels of α. The case where each level of β occurs with only one level of α is the arrangement called the "nested design." By contrast, the usual two-way design where every level of β occurs with a level of α is termed a "crossed design." Fig. 6.3 shows an example of a nested design.

Because the number of regions as well as the size of the respective regions in any class of the iteration is variable, the appropriate model discussed here is the unequal nesting, unequal number of replicates, nested two-factor design. To determine the test statistic, we derive the sum of squares under the alternative and the hypotheses H_a all $\alpha_i = 0$, H_b all $\beta_{j(i)} = 0$.

Class	Quadrants		
C_1	$R_{11}\ R_{12}\ \ldots\ R_{1,J_1}$		
C_2		$R_{21}\ R_{22}\ \ldots\ R_{2,J_2}$	
$\vdots$			
C_I			$R_{I1}\ R_{I2}\ \ldots\ R_{I,J_I}$

Figure 6.3. Two-way nested design. Classes are denoted by $C_i i = 1, 2, \ldots, I$. For each class C_i, there are $R_{i1}, \ldots, R_{i,J_i}$ corresponding quadrants. Each quadrant (i, j) is formed by K_{ij} pixels.

The sum of squares $SS_e(\mathbf{y}, \boldsymbol{\beta})$ is given by

$$SS_e(\mathbf{y}, \boldsymbol{\beta}) = \sum_{i=1}^{m} \sum_{j=1}^{J_i} \sum_{k=1}^{K_{ij}} (y_{ijk} - \mu - \alpha_i - \beta_{j(i)})^2 \tag{6.2}$$

Let the true cell mean v_{ij} be given by

$$v_{ij} = \mu + \alpha_i - \beta_{j(i)} \tag{6.3}$$

We can then rewrite Eq. (6.2) as

$$SS_e(\mathbf{y}, \beta) = \sum_{i=1}^{m} \sum_{j=1}^{J_i} \sum_{k=1}^{K_{ij}} (y_{ijk} - v_{ij})^2 \tag{6.4}$$

Therefore, the LSE of v_{ij} is obtained by minimizing Eq. (6.4) with respect to v_{ij}. Consequently, we obtain,

$$\widehat{v}_{ij} = y_{ij.} \tag{6.5}$$

The minimum of the sum of squares under the alternative is therefore equal to

$$SS_e(\mathbf{y}, \beta) = \sum_{i=1}^{m} \sum_{j=1}^{J_i} \sum_{k=1}^{K_{ij}} (y_{ijk} - y_{ij.})^2 \tag{6.6}$$

The class and segment homogeneity are based on the hypotheses H_a and H_b, respectively. The sum of squares for each case is derived in the usual manner, albeit in the case of the row effect, an alternate approach is used in the derivation of the estimates (see Appendix to Chapter 6). In this case, the sum of squares under H_a and H_b results in

$$SS_a(\mathbf{y}, \beta) - SS_e(\mathbf{y}, \beta) = \sum_{i=1}^{m} \sum_{j=1}^{J_i} \sum_{k=1}^{K_{ij}} (y_{ijk} - A_i)^2 \tag{6.7}$$

and

$$SS_b(\mathbf{y}, \beta) - SS_e(\mathbf{y}, \beta) = \sum_{i=1}^{m} \sum_{j=1}^{J_i} \sum_{k=1}^{K_{ij}} (y_{ij.} + y_{i..})^2 \tag{6.8}$$

respectively. The degrees of freedom associated with Eqs. (6.6)–(6.8) are found using the following. There are $\sum_{ij} K_{ij}$ observations in the design. Since there are $(m-1)$ independent row effects and $\sum_i (J_i - 1)$ column effects, the degree of freedom associated with $SS_e(\mathbf{y}, \beta)$ is

$$n_e = \sum_{ij} K_{ij} - (m-1) - \sum_i (J_i - 1) - 1 \tag{6.9}$$

The degrees of freedom corresponding to Eqs. (6.7) and (6.8) are

$$n_a = (m-1) \tag{6.10}$$

and

$$n_b = \sum_i (J_i - 1) \tag{6.11}$$

The test statistics for testing H_a and H_b are then

$$F_{class} = \frac{\sum_{i=1}^m \sum_{j=1}^{J_i} \sum_{k=1}^{K_{ij}} (y_{ijk} - A_i)^2}{\sum_{i=1}^m \sum_{j=1}^{J_i} \sum_{k=1}^{K_{ij}} (y_{ijk} - y_{ij.})^2} \cdot \frac{n_e}{n_a} \tag{6.12}$$

and

$$F_{segment} = \frac{\sum_{i=1}^m \sum_{j=1}^{J_i} \sum_{k=1}^{K_{ij}} (y_{ij.} - y_{i..})^2}{\sum_{i=1}^m \sum_{j=1}^{J_i} \sum_{k=1}^{K_{ij}} (y_{ijk} - y_{ij.})^2} \cdot \frac{n_e}{n_b} \tag{6.13}$$

The decision rules in this case are

$$\text{if } F_{class} > F_{\alpha,n_a,n_e} \quad \text{reject } H_a \tag{6.14}$$

Similarly,

$$\text{if } F_{segment} > F_{\alpha,n_b,n_e} \quad \text{reject } H_b \tag{6.15}$$

6.4 Logical predicates

The entire image is viewed as being formed of a number of classes, each class being split into a variable number of quadrants. What is important in this case is the representation of the partition in terms of a hierarchical model. The nested design lends itself naturally to this representation. The intermediate structure considered in the split-and-merge procedure is the tile. Each tile is split into four regions when the test of homogeneity is rejected. The logical step that is undertaken at the next stage is to divide each region into four quadrants, which are the primitive structures in the partition. This forms the splitting stage of the segmentation. An additional step is the location of the quadrants that are not homogeneous with the other components of the region under consideration. Therefore, two homogeneity tests are needed during this stage. We define the following logical predicates for the measure of homogeneity.

1. Predicate for tile homogeneity
At each level of the segmentation, it is required to test whether a set of pixels forming a tile is homogeneous in terms of gray level distribution. Let T_i denote the set. The predicate for homogeneity measure is defined as follows

$$P_T(T_I) = \begin{cases} \textbf{true} & \text{if } F_c < F_c^\alpha \text{ and } F_r < F_r^\alpha \\ \textbf{false} & \text{otherwise} \end{cases} \tag{6.16}$$

where F_c and F_r are the row and column test statistics corresponding to a two-way design ANOVA model, respectively. The threshold values F_c^α and F_r^α are tabulated in terms of the tile size M.

2. Predicate for region homogeneity

During the splitting procedure, a test of region homogeneity is carried out at each level of the partition. A number of homogeneity criteria have been developed and described elsewhere in the literature.[46] Among the well-known are those based on statistical tests that are defined in terms of the mean of the region under consideration.

The test of region homogeneity used here, which is based on row and column statistics, is identical to the tile homogeneity test. The region is declared homogeneous if $F_r < F_{\alpha,(m-1),(m-1)(m-1)}$ and $F_c < F_{\alpha,(m-1),(m-1)(m-1)}$. The data model in this case is the two-way model where the layout is of size $m \times m$. If the tests are rejected, that is, one of the statistics exceeds the threshold, the region is declared heterogenoues and split into four quadrants. Next, a test is performed to determine which of the quadrants is not homogeneous with the the other quadrants forming the specific region under test. Only adjacent quadrants are tested, which implies that at most only four tests are needed to determine the heterogeneous quadrants. Two logical predicates that measure the homogeneity of two adjacent quadrants are discussed next.

- Predicate based on F-statistics

  The observations from the two quadrants are combined to form a two-way layout. This is accomplished by connecting side-by-side the two quadrants. At this stage, we can perform a homogeneity test on the resulting data array using the row and column statistics described above. Let Q_i and Q_j, $i = 1, \ldots, 4; j = 1, \ldots, 4$ and $i \neq j$ be two adjacent quadrants.

$$P_{QF}(Q_i \cup Q_j) = \begin{cases} \textbf{true} & \text{if } F_c < F_c^\alpha \text{ and } F_r < F_r^\alpha \\ \textbf{false} & \text{otherwise} \end{cases} \quad (6.17)$$

  If the statistics do not exceed the thresholds, the quadrants are similar in structure and should be merged. Otherwise, they are heterogeneous. The data structure adopted in the merging strategy is based on linked lists. Each list in the partition is a class by itself and is formed of homogeneous and adjacent quadrants. Therefore, each time a quadrant is determined to be heterogeneous and not compatible with any quadrants of the same region, it is assigned to a new list. As shown in Fig. 6.2, a completely homogeneous region results in a single list, while four completely heterogeneous quadrants of the same region result in four linked lists.

- Predicate based on contrast functions

  A different logical predicate is defined based on the usual contrast test. Instead of calculating two F-statistics, a single contrast is computed for each valid combination of adjacent quadrants. Fig. 6.4 shows the principle behind the contrast test.

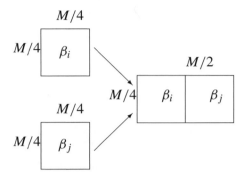

Figure 6.4. Contrast testing of quadrants.

Denote the column effects in Q_i and Q_j by β_i and β_j. Then, the contrast function is defined in terms of the column effects by

$$\psi = \sum_{i=1}^{M/4} \beta_i - \sum_{j=1}^{M/4} \beta_j \tag{6.18}$$

and the estimate is given by

$$\widehat{\psi} = \sum_{i=1}^{M/4} \widehat{\beta}_i - \sum_{j=1}^{M/4} \widehat{\beta}_j \tag{6.19}$$

The quadrants are declared homogeneous if $H_a : \psi = 0$ holds. Consequently, the predicate is defined as

$$P_{QC}(Q_i \cup Q_j) = \begin{cases} \textbf{true} & \text{if } |\widehat{\psi}| < (S F_{\alpha,1,\nu_e})^{1/2} \\ \textbf{false} & \text{otherwise} \end{cases} \tag{6.20}$$

where $S = \sigma_{\widehat{\psi}}^2$, and ν_e is the degree of freedom under Ω.

3. Predicate for segment homogeneity in the merging process

Fig. 6.5 shows an example of a partition corresponding to a segmented tile after region and quadrant homogenity tests were performed. The next stage is iterative in nature. Let P^j denote the partition at the jth level of the iteration and S_l and S_k two segments from the partition. The logical predicate for the measure of segment homogeneity is defined as follows

$$P_S(S_l \cup S_k) = \begin{cases} \textbf{true} & \text{if } F_{class} > F_C^\alpha \text{ and } F_{segment} < F_S^\alpha \\ \textbf{false} & \text{otherwise} \end{cases} \qquad (6.21)$$

where F_{class} and $F_{segment}$ are the row and column test statistics corresponding to a fixed-effects nested design model.

Region II	S_1	04	06	26
Region II	S_2	24		
Region III	S_3	44	64	
Region III	S_4	46	66	
Region IV	S_5	40	62	60
Region IV	S_6	42		
Region I	S_7	00		

Figure 6.5. Initial partition of tile with size 8×8. Each quadrant is indexed by its horizontal and vertical coordinates.

6.5 Adaptive class formation

The splitting procedure results in the formation of linked lists for each of the four regions of the tile. For a given size M of the tile, there are at most 16 linked lists when all quadrants are heterogeneous and four lists when all the regions are homogeneous. The merging process consists first in merging adjacent quadrants from each region. Then, a nested design is performed by including adjacent quadrants from neighboring regions. The implementation of the merging procedure is as follows.

Region I	Region II
Region IV	Region III

Figure 6.6. Region and quadrant labeling for merging process.

Denote each region and its associated linked lists by R_i, $i = 1, 2, 3, 4$ and L_{ij}, where j varies between 1 and 4 depending on the number of homogeneous quadrants in the region. A graphic representation is shown in Fig. 6.6. The merging is started using the region with at least two segments (see Section 5.1) as the reference region. For the example in Fig. 6.8, the process is started with region II because region I is homogeneous and consists of only one segment. At each stage, one linked list, that is, segment from the other regions, is compared to the segments of the reference region. Adjacency is checked before any merging can occur. In the present structure, the nested design is used for the data parameterization. As should be clearly evident at this stage, the nested design representation is a natural one and fits the objectives of the merging process very well. Fig. 6.7 shows the first design obtained by using the linked lists in region II as nested region effects in addition to the linked list of region I. In the example, we have five quadrants in the design with two classes. Based on the analysis developed in Section 6.3, the region statistic $F_{segment}$ determines whether the new quadrants in the new list are homogeneous with those in the initial list. At the same time, F_{class} provides a way to check whether the two classes are still disjoints after the inclusion of the new quadrants.

Class	Quadrants				
C_1	04	06	26	00	
C_2					24

Figure 6.7. Starting configuration of the nested design using S_1, S_2 and S_7.

6.5.1 Test outcomes

Four possible cases are considered depending on the segment and class test statistics.

(1) $F_{class} > F_{\alpha,n_a,n_e}$ and $F_{segment} > F_{\alpha,n_b,n_e}$.
Because the segment test is rejected, the new segments are not homogeneous with those in the list. Note that here the class test is also rejected, which means that the classes are still disjoints. However, this is of no consequence because the segment test was rejected. The next step in the class formation is to move the quadrants to the linked list following the previous list. If all lists are exhausted without a match, the quadrants forming the segment under test are considered as a new class and therefore have to be included in the original nested design. The class size is consequently increased by one.

(2) $F_{class} < F_{\alpha,n_a,n_e}$ and $F_{segment} > F_{\alpha,n_b,n_e}$.
Ideally, this case should not occur. The fact that the class test is accepted implies that the classes are homogeneous and therefore should be merged. However, the regions in at least one class, that is, linked list in the design after the inclusion of the new segment, are not homogeneous. This, therefore, contradicts the assumption

that the tile has been partitioned into homogeneous regions. When this case occurs, the threshold values used in the initial tests of row and column statistics should be changed for higher values of the confidence level α.

(3) $F_{class} < F_{\alpha,n_a,n_e}$ and $F_{segment} < F_{\alpha,n_b,n_e}$.
Again, this case reflects the fact that the regions together with the classes of the partition are homogeneous, which contradicts the results of the tile partition.

(4) $F_{class} > F_{\alpha,n_a,n_e}$ and $F_{segment} < F_{\alpha,n_b,n_e}$.
Whenever we have the present case, the new quadrants in the segment under test are compatible with one specific segment in the design. Therefore, list L_{ij} is updated by including the new quadrants of the segment under test. Next, the remaining lists from the initial partition are tested.

An illustrative example
Fig. 6.8 shows an 8×8 tile with three distinct gray levels. The region and quadrant testing results in the partition shown in Fig. 6.5. This is the initial partition that serves as the input for the iterative nested design. Table 6.1 shows the sequence of segments merges. Also shown are the values of segment and class statistics with the corresponding degrees of freedoms.

189	194	187	187	195	188	188	196
189	188	187	193	188	188	194	189
189	195	187	186	19	21	189	187
192	187	188	193	18	16	187	190
50	48	20	19	49	51	190	184
48	49	21	16	50	52	187	190
50	44	47	50	58	48	188	193
49	49	51	42	48	48	197	186

Figure 6.8. 8×8 tile with three distinct gray level values.

In this example only cases I and IV of the test outcomes are generated. As an example of the segment merges, consider the fourth iteration. Because $F_S > F_S^\alpha$ and $F_R > F_R^\alpha$, segment S_5 is not merged with S_8. Therefore, the next iteration involves the second neighbor of S_5, which is S_3. In this case the statistics verify the conditions of case IV, which implies that the two segments can be merged to form segment S_9. Fig. 6.9 shows a graphic representation of the segment merges.

Table 6.1. *Statistics for the partition in Fig. 6.8.*

Partition	Statistics	Thresholds	Outcome
S_1 $S_7 = S_8$ S_2 S_3 S_4 S_5 S_6	$F_{class} = 6467.01$ $F_{segment} = .739$	$F_{5,51} = 2.40$ $F_{7,51} = 2.20$	Merge
S_8 S_6 S_2 S_3 S_4 S_5	$F_{class} = 5301.71$ $F_{segment} = 1306.11$	$F_{4,51} = 2.56$ $F_{8,51} = 2.13$	No merging
S_8 S_2 S_3 S_6 S_4 S_5	$F_{class} = 8015.97$ $F_{segment} = 34.54$	$F_{4,51} = 2.56$ $F_{8,51} = 2.13$	No merging
S_8 S_5 S_2 S_3 S_4 S_6	$F_{class} = 4014.94$ $F_{segment} = 2156.8$	$F_{4,51} = 2.56$ $F_{8,51} = 2.13$	No merging
S_8 S_2 S_3 $S_5 = S_9$ S_4 S_6	$F_{class} = 8083.04$ $F_{segment} = 1.00$	$F_{4,51} = 2.56$ $F_{8,51} = 2.13$	Merge
S_8 $S_4 = S_{10}$ S_2 S_3 S_5 S_6	$F_{class} = 10723.72$ $F_{segment} = .899$	$F_{3,51} = 2.79$ $F_{9,51} = 2.07$	Merge
S_{10} S_2 S_9 S_6	$F_{class} = 15926.19$ $F_{segment} = 32.68$	$F_{2,51} = 3.18$ $F_{10,51} = 2.02$	No merging

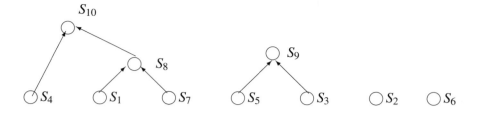

Figure 6.9. Graphic representation of segment merges.

6.5.2 Merging strategies

The linking of the quadrants from adjacent tiles, in the vertical and horizontal directions, is performed in three steps. First quadrants in a tile are merged using the adaptive nested design strategy. Second is the horizontal linking that connects quadrants from adjacent overlapping tiles. Finally, column merging connects quadrants from vertically overlapping tiles. The present strategy is similar to the tile-by-tile approach proposed in reference [48].

Tile merging

Lists from three regions in the tile are tested against those homogeneous lists in a reference region. Ideally, the reference region is region I, and the merging is started from region IV. In the actual implementation, we consider two extreme cases. First, if region I is homogeneous, it implies that the degree of freedom n_a corresponding to class effects is identically equal to zero. Consequently, another region that is formed of at least two linked lists, which guarantees a nonzero degree of freedom, is chosen as the reference region against which the three remaining regions are compared. The second extreme case happens when the tile is partitioned into four homogeneous regions, therefore leading to exactly four linked lists. In this case the degree of freedom corresponding to region effects is identically equal to zero. As a result, without resorting to nested design tests, we consider the tile to be formed of four lists.

Column merging

In the present implementation the image is scanned top to bottom starting from the left top corner. At each scan the mask is moved by $M/2$ pixels so that the $M/2$ rightmost columns from the previous tile are included in the new tile. Fig. 6.10 shows how regions from scan A are connected with regions from scan B. The reason behind this scheme is to allow a merging of the new lists with those obtained in the previous scan using regions A.II and A.III as a junction between regions A.I–A.IV and B.II–B.III. The same strategy is adopted along each row in the image.

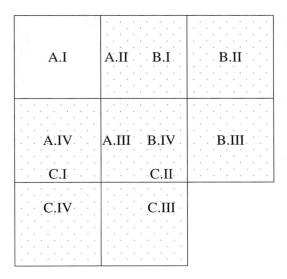

Figure 6.10. Overlapping tiles in row and column scans. The row junction is represented by regions {A.II–B.I and A.III–B.IV} while the column junction is represented by {A.III–C.II and A.IV–C.I}, respectively.

Region merging

After each row is processed, the resulting lists are joined with those obtained in the previous scans. Again, the same strategy used in row merging is applied in column merging. That is, after two successive row mergings are obtained, the segments in both columns that have identical quadrants are linked together.

6.6 Concluding remarks

A procedure for the segmentation of gray level images is developed using a nested design algorithm. The specific model considered is based on a fixed effects parameterization for gray level images. Starting from the top left corner, the procedure initially splits overlapping square tiles into homogeneous regions. This results in linked lists of similar quadrants. As a result, a nested design is formed with linked lists representing row effects, and quadrants in a list form column effects. Using row and column statistics, that is, class and segment statistics, homogeneity is determined by comparing the statistics to tabulated thresholds.

Linked lists from adjacent lists are successively added to the design until all segments are exhausted. Because of the localized nature of the procedure, we consider row and column merging in linking regions from adjacent tiles.

Several logical predicates are defined in this chapter. Starting with a tile, a

<table>
<tr><td>(a)</td><td>(b)</td></tr>
</table>

Figure 6.11. "Lena" image. clean image.

Figure 6.12. Segmentation of "Lena" image. $\alpha = 5\%$ for all decision cases.

predicate for homogeneity is defined in terms of corresponding row and column statistics. A similar predicate is defined for regions. We also consider a predicate defined in terms of a contrast function for locating those quadrants that are not uniform with the other constituent quadrants of a nonuniform region. Finally, a predicate for segment merging is defined in terms of the class and segment statistics of the nested design.

A typical result for the segmentation process is demonstrated in Figs. 6.11 and 6.12.

Appendix

To determine the sum of squares under H_a; all $\alpha_i = 0$, the same approach used for the two-way design with unequal number of observations in the cells is implemented (see reference 2, p. 116).

The interpretation given to the sum of squares $SS_A = SS_a(\mathbf{y}, \beta) - SS_e(\mathbf{y}, \beta)$ in the numerator of the F-statistic is that it is equal to the square of the maximum estimated normalized contrast in the main effects α_i, that is, $SS_A = \psi_{max}^2$. Let the main row effect α_i be defined as follows

$$\alpha_i = A_i - \mu \tag{A.1}$$

where $A_i = \sum_j n_{ij}$ and $\mu = \sum_i \sum_j n_{ij}$.

The maximum contrast $\widehat{\psi}$ is found by maximizing $\widehat{\psi} = \sum_i c_i \widehat{A}_i$ subject to the constraints

$$\sum_i c_i = 0 \tag{A.2.a}$$

and

$$\sum_i \sum_j \frac{c_i^2}{K_{ij}} = 1 \tag{A.2.b}$$

Let $\Delta = \sum_i c_i \widehat{A}_i - \lambda_1 \sum_i c_i - \lambda_2 \sum_i \sum_j \frac{c_i^2}{K_{ij}}$. The objective here is to find the coefficient set c_i that maximizes the estimated contrast. Therefore, by taking the derivative with respect to c_i, we obtain

$$\frac{\partial \Delta}{\partial c_i} = \widehat{A}_i - \lambda_1 - 2\lambda_2 \sum_j \frac{c_i}{K_{ij}} = 0 \tag{A.3}$$

or

$$c_i = \frac{\widehat{A}_i - \lambda_1}{2\lambda_2} \left(\sum_j \frac{1}{K_{ij}} \right)^{-1} = \frac{\widehat{A}_i - \lambda_1}{2\lambda_2} W_i \tag{A.4}$$

where $W_i = \left(\sum_j \frac{1}{K_{ij}} \right)^{-1}$.

Consider the first condition $\sum_i c_i = 0$. Using Eq. (A.5), we obtain

$$\frac{1}{2\lambda_2} \sum_i (\widehat{A}_i - \lambda_1) W_i = 0 \tag{A.5}$$

Similarly, using Eq. (A.3), we obtain

$$\sum_i \frac{c_i^2}{W_i} = \sum_i \frac{(\widehat{A}_i - \lambda_1)^2 W_i}{4\lambda_2^2} = 1 \tag{A.6}$$

which implies that

$$\lambda_2^2 = \sum_i \frac{(\widehat{A}_i - \lambda_1)^2 W_i}{4} \tag{A.7}$$

From Eq. (A.6) we have

$$\lambda_1 = \sum_i \frac{\widehat{A}_i W_i}{\sum_i W_i} \tag{A.8}$$

Finally, from the definition of SS_A, we have

$$SS_A = SS_a(\mathbf{y}, \boldsymbol{\beta}) - SS_e(\mathbf{y}, \boldsymbol{\beta}) = \left(\sum_i c_i \widehat{A}_i \right)^2 = \sum_i (\widehat{A}_i - \lambda_1)^2 W_i \tag{A.9}$$

7

Radial masks in line and edge detection

7.1 Introductory remarks

In this chapter the methodology presented in Chapters 3 and 4 is extended to include processing of images using radial masks. The approach produces higher power (improved detectability) in most image processing operations at a small cost of increase in processing time. Also, the radial processing masks are less sensitive to the degree of correlation of the background noise.

Because of the similarity of some of the mathematical developments, many of the details in describing radial mask operations are omitted. The analysis involves the Markov noise model with the general results easily reduced to the independent noise case by replacing the Markov dependence covariance matrix with a diagonal matrix.

In the first part of the chapter, the potential features (line or edge elements) are extracted using a radial version of the masks for the one-way designs. Next, the symmetrical incomplete block design (SBIB) technique is generalized to include radial processing. This results in improvement in the feature extraction process for a fixed alarm rate.

The contrast function approach is also extended to include radial masking techniques. The algorithm is capable of detecting potential features and their locations simultaneously. The decision threshold is determined by the variance of the contrast function and the correlation coefficient of the noise. All three procedures perform well in dependent noise; they are simple, fast, and relatively robust.[66]

7.2 Radial masks in one-way ANOVA design

For a noisy image contaminated by correlated Gaussian noise, the one-way ANOVA[2] is given by

$$\mathbf{y} = X^T \beta + \mathbf{e} \tag{7.1}$$

where β is an unknown parameter vector, rank$(X) = n$ and $E(\mathbf{ee}^T) = \sigma^2 \mathbf{K}_f$.

147

Using a prewhitening transformation (see Chapter 2), and proceeding with the minimization of the error sum of squares under the alternative Ω, we obtain the system of equations

$$\sum_{i=1}^{m}\sum_{j=1}^{n} R_{.kij}\,y_{ij} = R_{.k..}\mu + \sum_{j=1}^{n} R_{.k.j}\beta_j \qquad k = 1, 2, \ldots, n \tag{7.2}$$

where $R_{..kl} = \sum_{i=1}^{m}\sum_{j=1}^{n} R_{ijkl}$. Furthermore, for the general mean μ, we have

$$\sum_{i=1}^{m}\sum_{j=1}^{n} R_{..ij}\,y_{ij} = R_{....}\mu + \sum_{j=1}^{n} R_{...j}\beta_j \tag{7.3}$$

Finally, we can write Eqs. (7.2)–(7.3) in matrix form

$$\begin{pmatrix} R_{.1.1} & R_{.1.2} & \cdots & R_{.1.n} & R_{.1..} \\ R_{.2.1} & R_{.2.2} & \cdots & R_{.2.n} & R_{.2..} \\ \cdots & \cdots & \cdots & \cdots & \cdots \\ R_{.n.1} & R_{.n.2} & \cdots & R_{.n.n} & R_{.n..} \\ R_{...1} & R_{...2} & \cdots & R_{...n} & R_{....} \end{pmatrix} \begin{pmatrix} \beta_1 \\ \beta_2 \\ \cdots \\ \beta_n \\ \mu \end{pmatrix} = \begin{pmatrix} R_{.111} & \cdots & R_{.1mn} \\ R_{.211} & \cdots & R_{.2mn} \\ \cdots & \cdots & \cdots \\ R_{.n11} & \cdots & R_{.nmn} \\ R_{..11} & \cdots & R_{..mn} \end{pmatrix} \begin{pmatrix} y_{11} \\ y_{12} \\ y_{13} \\ \cdots \\ y_{mn} \end{pmatrix} \tag{7.4}$$

or, equivalently,

$$\mathbf{W}\beta = \mathbf{Z}\mathbf{y} \tag{7.5}$$

As usual, the matrix of effects, $\mathbf{E} = \mathbf{W}^{-1}\mathbf{Z}$, is calculated once for the whole process, and the LSE of parameters, $\widehat{\beta} = \mathbf{E}\mathbf{y}$, is obtained each time after the data collection.

It can be shown that the error sum of squares is given by

$$SS_e = (\mathbf{Q}\mathbf{y})^T \mathbf{P}\mathbf{P}^T (\mathbf{Q}\mathbf{y}) = (\mathbf{Q}\mathbf{y})^T \mathbf{K}_f^{-1}(\mathbf{Q}\mathbf{y}) \tag{7.6}$$

where $\mathbf{P}\mathbf{P}^T = \mathbf{K}_f^{-1}$ and $\mathbf{Q} = \mathbf{I} - \mathbf{X}^T\mathbf{E}$. Under the hypothesis, that is, all $\beta_j = 0$, the general mean estimate $\widehat{\mu}_a$ is given by

$$\widehat{\mu}_a = \sum_{i=1}^{m}\sum_{j=1}^{n} R_{..ij}\,y_{ij} \tag{7.7}$$

and the numerator of the F-statistic may be written as

$$SS_a - SS_e = \mathbf{Q}_a^T \mathbf{K}_f^{-1}\mathbf{Q}_a \tag{7.8}$$

where $\mathbf{Q}_a = \widehat{\mu}_a - \mu_\Omega - \beta$. Thus, by replacing Eqs. (7.6) and (7.8) in the F-statistic formulation, we obtain

$$F = \frac{n(m-1)}{(n-1)} \cdot \frac{\mathbf{Q}_a^T \mathbf{K}_f^{-1}\mathbf{Q}_a}{(\mathbf{Q}\mathbf{y})^T \mathbf{K}_f^{-1}(\mathbf{Q}\mathbf{y})} \tag{7.9}$$

Eq. (7.9) is used in formulating the radial version of the one-way model in the detection and extraction of edges and lines.

7.3 Boundary detection procedure

The radial mask operator developed in the preceding section is applied to image data contaminated by dependent noise. Observation samples are originated at the given radial mask with an arbitrary direction θ. Let T_θ denote a set of masks. The neighboring pixels of T_{θ_i} depend on the directional parameter θ_i, as is shown in Fig. 7.1.

The F-test setting of Section 3.7 is closely related to the approach used in radial processing. That is, we are mainly interested in block effects within the mask. The potential boundary elements of interest are related to the highest value of F-statistic in T_θ.

In a SBIB design, the hypothesis test is carried out simultaneously on the block mean using each block as a basis for the multiple comparisons (see Chapters 3 and 4).

Even though the F-statistic confirms the potential boundary elements, it cannot give the location of these elements. As in Chapter 3, the shape test is required to locate the appropriate elements that belong to edges or lines.

The F-statistic of a SBIB design used in the detection of a trajectory or line in case of independent noise has been used in reference [39] by comparing it with the threshold corresponding to a given confidence level; namely, by thresholding the adjusted block mean by the total mean. The unbiased estimate of the block mean b_j in the SBIB design for the two-way ANOVA[14] is

$$b_j = \frac{1}{\lambda t} \left[K\beta_j - \sum_{i=1}^{m} T_{ij} d_{ij} \right] + \frac{G}{n} \qquad j = 1, 2, \ldots, m. \tag{7.10}$$

In the one-way ANOVA case, the block effects are given by

$$\widehat{\beta}_j = b_j - \mu = y_{.j} - \widehat{\mu} \tag{7.11}$$

From Eqs. (7.10) and (7.11), the arithmetic mean of block mean b_j is then equal to

$$b_n = \frac{1}{m} \sum_{i=1}^{m} b_j \tag{7.12}$$

For the dependent noise case, the block mean estimate is obtained from $\beta = \mathbf{E}\mathbf{y}$ because the adjusted block mean terms are meaningless. The matrix of effects $\mathbf{E}$ is calculated once from the known correlation coefficients or their estimate. The observation samples $\mathbf{y}$ are in this case given by the radial mask T_{θ_i}.

The threshold T is defined by

$$T = U(\widehat{\beta}_n - \widehat{\beta}_j) \tag{7.13}$$

where U is the unit step function, β_n is the estimate of the mean block effects, and $\widehat{\beta}_j$ is the estimate of block effects. By using the threshold function, the sequences of 1s and 0s are stored to determine the potential boundary elements. If these sequences belong to the class of boundary elements, one can locate them at the proper place where the highest F-value in T_θ is given by the mask operation.

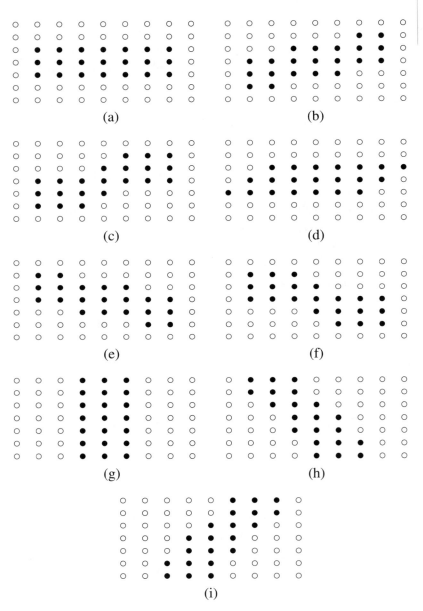

Figure 7.1. Radial mask structures.

7.4 Contrast-based detectors using radial masks

The theory behind the use of contrast-based detectors using radial masks is the same as that presented in Chapters 3 and 4. The only difference is that the radial masks of Fig. 7.1 are used to scan the image.

In the following, we assume that the boundary elements (edges and lines) of interest may be among the patterns shown in Figs. 7.2 and 7.3. The block estimate of the sample means is the average of gray levels of all block pixels, or

$$\widehat{\beta}_j = \frac{1}{m} \sum_{i=1}^{m} y_{ij} \tag{7.14}$$

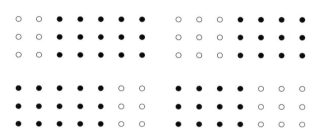

Figure 7.2. Potential edge structures.

If we consider that edges are to be detected, then we may use the following set of contrast functions, each one being calibrated to a particular shape. The estimate of the contrast function for each shape is then given by (see Fig. 7.2)

$$\widehat{\psi}_1 = \tfrac{2}{7}(\widehat{\beta}_3 + \widehat{\beta}_4 + \widehat{\beta}_5 + \widehat{\beta}_6 + \widehat{\beta}_7) - \tfrac{5}{7}(\widehat{\beta}_1 + \widehat{\beta}_2) \tag{7.15}$$

$$\widehat{\psi}_2 = \tfrac{2}{7}(\widehat{\beta}_1 + \widehat{\beta}_2 + \widehat{\beta}_3 + \widehat{\beta}_4 + \widehat{\beta}_5) - \tfrac{5}{7}(\widehat{\beta}_6 + \widehat{\beta}_7) \tag{7.16}$$

$$\widehat{\psi}_3 = \tfrac{3}{7}(\widehat{\beta}_4 + \widehat{\beta}_5 + \widehat{\beta}_6 + \widehat{\beta}_7) - \tfrac{4}{7}(\widehat{\beta}_1 + \widehat{\beta}_2 + \widehat{\beta}_3) \tag{7.17}$$

$$\widehat{\psi}_4 = \tfrac{3}{7}(\widehat{\beta}_1 + \widehat{\beta}_2 + \widehat{\beta}_3 + \widehat{\beta}_4) - \tfrac{4}{7}(\widehat{\beta}_5 + \widehat{\beta}_6 + \widehat{\beta}_7) \tag{7.18}$$

Figure 7.3. Potential line structures.

For the three-pixel line, the contrast functions (see Fig. 7.3) can be written as

$$\widehat{\psi}_5 = \tfrac{4}{7}(\widehat{\beta}_2 + \widehat{\beta}_3 + \widehat{\beta}_4) - \tfrac{3}{7}(\widehat{\beta}_1 + \widehat{\beta}_5 + \widehat{\beta}_6 + \widehat{\beta}_7) \tag{7.19}$$

$$\widehat{\psi}_6 = \tfrac{4}{7}(\widehat{\beta}_3 + \widehat{\beta}_4 + \widehat{\beta}_5) - \tfrac{3}{7}(\widehat{\beta}_1 + \widehat{\beta}_2 + \widehat{\beta}_6 + \widehat{\beta}_7) \qquad (7.20)$$

$$\widehat{\psi}_7 = \tfrac{4}{7}(\widehat{\beta}_4 + \widehat{\beta}_5 + \widehat{\beta}_6) - \tfrac{3}{7}(\widehat{\beta}_1 + \widehat{\beta}_2 + \widehat{\beta}_3 + \widehat{\beta}_7) \qquad (7.21)$$

As seen from the above equations, the contrast functions depend on the edge or line patterns and the radial mask designs, the latter by virtue of the pixel location as determined by the specific mask.

First, we select $\widehat{\psi}_{max}$ corresponding to an important potential element. The location of the potential element is determined by the variance $\widehat{\psi}_{max}$ and the pattern of these elements. Since the correlation coefficient matrix corresponding to the radial mask type is known, the variance of $\widehat{\psi}_{max}$ is given by

$$\widehat{\sigma}_{\widehat{\psi}}^2 = s^2 \mathbf{a}_{max}^T \mathbf{K}_f^{-1} \mathbf{a}_{max} \qquad (7.22)$$

which is a generalization of Eq. (3.28) for dependent noise, and $\mathbf{a}_{max}$ is a vector of coefficients that correspond to the pattern of $\widehat{\psi}_{max}$.

For instance, let $\max(\widehat{\psi}) = \widehat{\psi}_2$,

$$\widehat{\psi}_2 = \frac{2}{7m} \sum_{i=1}^{m} (y_{i1} + y_{i2} + y_{i3} + y_{i4} + y_{i5}) - \frac{5}{7m} \sum_{i=1}^{m} (y_{i6} + y_{i7}) \qquad (7.23)$$

Hence, by inspection (with $m = 3$),

$$\mathbf{a}_2^T = [(2/21\ 2/21\ 2/21 \ldots -5/21\ -5/21\ -5/21| \ldots | 2/21\ 2/21$$
$$\ldots -5/21)] \qquad (7.24)$$

This leads to $s^2 = \tfrac{1}{18}(\mathbf{Zy})^T \mathbf{K}_f^{-1}(\mathbf{Zy})$ and $T_0 = (2F_{\alpha,2,8})^{1/2} \widehat{\sigma}_{\widehat{\psi}_2}$ where T_0 is the threshold for $\widehat{\psi}_2$, and $\mathbf{Z} = \mathbf{I} - \mathbf{X}^T \mathbf{E}$.

The projection matrix $\mathbf{Z}$ projects the vector $\mathbf{y}$ on the orthonormal complement of the space $\mathbf{V}$, which expands the mean vector $\mathbf{X}^T \beta$ of the vector $\mathbf{y}$.

In the final stage we compare $\widehat{\psi}_{max}$ to the predetermined threshold. If the hypothesis is rejected, we locate the potential element by the pattern of $\widehat{\psi}_{max}$; namely, the starting pixel of the potential edge in $\widehat{\psi}_{max}$ is allocated to the edge, and the center pixel assigned to the line.

7.5 Power calculation

The contrast function of the radial masks is used to find the block effects simultaneously by the multiple comparison S-method. The aim is to see the difference between the regular ANOVA mask structures and the radial structures in terms of the probability of detection.

The underlying assumptions are that the observation data $\mathbf{Y}$ are $N(\mathbf{X}^T \beta, \sigma^2 \mathbf{K}_f)$ and the contrast function is $\widehat{\psi} = \sum_{i=1}^{m} c_i \beta_i$. The probability of detection for all ψ

is given by

$$Pr(|\psi - \widehat{\psi}| \leq S\widehat{\sigma}_{\widehat{\psi}}) = 1 - \alpha \tag{7.25}$$

where $S^2 = q F_{\alpha;q;n-r}$ and $\widehat{\sigma}_{\widehat{\psi}}$ is the variance estimate of the contrast function $\widehat{\psi} \in L$. The F-test is equivalent to the rejection of the hypothesis H_0 if $|\widehat{\psi}| > S\widehat{\sigma}_{\widehat{\psi}}$; namely, the LSE estimate $\widehat{\psi}$ of ψ is significantly different from zero by the S-criterion (multicomparison method). The probability of rejecting the hypothesis H_0 is the same as the probability that the maximum of $\widehat{\psi}$ is greater than the threshold. Thus, the probability of detection, when some block effects exist within the radial masks, can be represented by

$$P_D = Pr(|\widehat{\psi}_{max}| > S\widehat{\sigma}_{\widehat{\psi}_{max}}|H_1) \tag{7.26}$$

We can see from Eqs. (7.25) and (7.26) that the Neyman–Pearson detector can be constructed. Under H_0, one of the contrast functions ψ_{max} is $N(0, \widehat{\sigma}_{\psi_{max}}^2)$, and under H_1 it is $N(\widehat{\psi}_{i_{max}}, \widehat{\sigma}_{\psi_{i_{max}}}^2)$, where $E(\widehat{\psi}_{i_{max}}) = \psi_{i_{max}}$.

The false alarm rate can be written as

$$P_f = Pr(|\widehat{\psi}_{max}| > S\widehat{\sigma}_{\widehat{\psi}_{max}}|H_0) \tag{7.27}$$

As in the Neyman–Pearson detector, we seek the highest probability of detection P_D for a given false alarm rate, that is, $P_f = \alpha$. Thus, the contrast-to-noise ratio (CNR) is defined as follows

$$CNR = \frac{E(\widehat{\psi})}{\widehat{\sigma}_{\widehat{\psi}}} \tag{7.28}$$

It is easy to see from the definition that the maximum estimate of the contrast function for a certain mask yields the highest power of detection, even though the variance of $\widehat{\psi}$ should be considered. In addition, P_D depends on the threshold value $S\widehat{\sigma}_{\widehat{\psi}_{max}}$. The contrast variance is an important factor in determining the probability of detection for the contrast algorithm.

0	0	d	d	d	d	d
0	0	d	d	d	d	d
0	0	d	d	d	d	d

Figure 7.4. Basic edge structure.

Consider the edge shown in Fig. 7.4 as a primitive feature to investigate the relationship between P_D and the correlation coefficients of the noise component. We concentrate on the contrast function ψ_1 to analyze the probability of detection.

The block estimates $\{\widehat{\beta}_i\}$ are the sample means of observation data through the columns. The estimate of the contrast ψ_1 yields

$$\widehat{\psi}_1 = \frac{2}{21}(y_{.3} + y_{.4} + y_{.5} + y_{.6}) - \frac{10}{21}(y_{.1} + y_{.2}) \tag{7.29}$$

where $y_{.j} = \sum_{i=1}^m y_{ij}$. It follows that $E(\widehat{\psi}_1)$ is equal to 10/7 with d = 1. The variance of the contrast function can be obtained by calculating the variance between blocks, after calculating the variance within block $y_{.j}$ for each j. In general, the correlation matrix of the underlying noise in radial masks using SBIB designs is generalized by the positions of the pixels. We assume that this matrix has a tensored form, $\mathbf{K}_f = \mathbf{K}_C \otimes \mathbf{K}_R$, to calculate the variance of $\widehat{\psi}$.

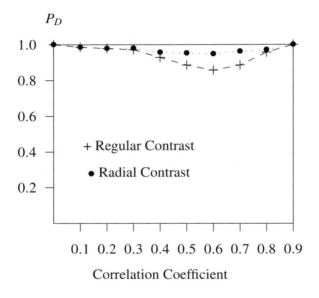

Figure 7.5. Regular vs radial mask detection.

Let a denote the row correlation coefficient and b denote the column correlation coefficient. The variance within the block is given by

$$\sigma_{y_{.j}}^2 = \sigma^2(3 + 4a + 2a^2) \tag{7.30}$$

After calculating the variance between blocks and rearranging the results, we have

$$\sigma_{\widehat{\psi}}^2 = \frac{\sigma^2}{49}(3 + 4a + 2a^2)(70 + 60b - 16b^2 - 24b^3 - 32b^4 - 40b^5 - 20b^6) \tag{7.31}$$

Since $\widehat{\psi}$ is normally distributed with mean ψ and variance $\widehat{\sigma}_{\widehat{\psi}}^2$, that is, $N(\psi, \widehat{\sigma}_{\widehat{\psi}}^2)$,

P_D can be written as

$$P_D = \frac{1}{\sqrt{2\pi}\sigma_{\widehat{\psi}}} \int_{S\sigma_{\widehat{\psi}}}^{\infty} \exp\frac{-Z^2}{2} dZ \tag{7.32}$$

where $S = 2F_{\alpha,2,8}$ and $Z = \frac{x-10/7}{\sigma}$. The probabilty of detection exhibits the V curve behavior as indicated before.

The probability of detection exhibits a convex form as a function of the correlation coefficients, with the minimum point falling between $a = 0.5$ and $a = 0.6$. Fig. 7.4 shows the variation of P_D versus the correlation coefficient a. Clearly, we see the improvement obtained in introducing radial masks in the design.

The level of threshold values is more likely to be adaptive under the dependent noise condition, whereas it is fixed regardless of noise conditions in the regular ANOVA contrast algorithm. Though the computing time for the radial mask is somewhat longer than for conventional contrast algorithms based on ANOVA techniques, its use is justified by the considerable improvement in performance. The regular contrast algorithm based on ANOVA techniques is still useful for low correlation coefficients or for high contrast-to-noise, since it is simple and fast. Moreover, it is applicable in real time processes. For scanning the image, we use a (9×9) block of pixels. Then, a radial mask is applied to select the appropriate pixel elements. The entire image is scanned every three columns and rows. The location of the potential elements is determined by the contrast function $\widehat{\psi}_{max}$, whereas it can be decided by the shape test in the modified ANOVA techniques.

7.6 Concluding remarks

In this chapter the theory of line and edge detectors is extended to include radial processing masks. The radial masking techniques in the presence of correlated noise are more sensitive to line and/or edge elements than the rectangular masks described in Chapters 3 and 4 (see Fig. 7.5). The improvement in performance of radial masks is also demonstrated for contrast-function-based detectors. The contrast algorithm calculates only one sum of squares of the F-statistic for a given threshold level. This algorithm adapts well to the noise conditions and to the feature patterns because the threshold level is determined by the variance of the contrast function and the correlation coefficient of noise. The contrast function algorithm using radial masks performs better than the conventional contrast algorithm.

Though the performance of the radial masks is better than that of the conventional masks, they all demonstrate a deterioration in performance for values of the correlation coefficient between 0.45–0.55. This deterioration can be partially explained by the fact that in this range of values, the noise background tends to generate some pseudo-features. The threshold level depends only on the feature patterns, neglecting the change of correlated noise conditions.

8

Performance analysis

8.1 Stochastic approximation in parameter estimation

8.1.1 Remarks

Consider the general linear model

$$\mathbf{y} = \mathbf{X}^T \boldsymbol{\beta} + \mathbf{e} \tag{8.1}$$

where $\mathbf{e}$ is $N(0, \sigma^2)$. The parameter estimate may be obtained using the least squares approach. Toward that end, we minimize the functional $\Lambda = (\mathbf{y} - \mathbf{X}^T \boldsymbol{\beta})^T (\mathbf{y} - \mathbf{X}^T \boldsymbol{\beta})$ for every unknown parameter $(\beta_1, \beta_2, \ldots, \beta_p)$ of the vector $\boldsymbol{\beta}$. The solution of the normal equation $\partial \Lambda / \partial \beta_j = 0$ yields the LSE estimator. Under the Gaussian assumption, this leads to the usual F-statistic. This approach is still valid if there are small to moderate deviations from the Gaussian distribution. This can be verified by extensive simulations. Many of them were performed by the authors and some of the senior author's doctoral students. Also, one should consult reference [79].

In the presence of impulsive noise, the LS approach is no longer applicable, and alternate techniques must be used. In this case, we consider two different situations. In the first case noise contamination is not severe, in which case the parameters of the linear model are estimated using the nonrobustized version of the Robbins–Monro stochastic approximation (RMSA) algorithm. Second, if the background noise is severe, we use the robustized version of the RMSA estimator. Though the Wilcoxon statistic preprocessor is discussed in detail, other approaches are possible.

For both cases the noise components are assumed to be independent identically distributed (i.i.d). For those readers interested in the noise-dependent case, a reference is given for a recent work that extends the approach to strongly mixing type of dependence.

8.1.2 Stochastic approximation versus classical estimation

Before considering a detailed analysis of the stochastic approximation theory, it is interesting to relate it to the classical estimation theory. Because of the widespread use of digital computers, only the discrete-time and discrete-parameter case will be presented.

In classical terms the estimation problem may be presented as shown in Fig. 8.1.

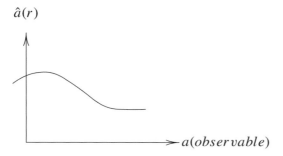

Figure 8.1. Classical estimation.

The two basic classical estimation approaches and their associated characteristics are as follows.

Bayes estimation

In this model, we assume we have (1) $p(\mathbf{a})$, (2) $p(\mathbf{r}/\mathbf{a})$, and (3) risk function. Based on this knowledge, we minimize the average risk. Special cases of the model are the following.

(1) Minimum mean-square estimator (MMSE)
 The estimator is $\hat{a}(\mathbf{r}) = E[\mathbf{a}/\mathbf{r}]$, which is the well-known regression function. If the distributions are Gaussian, $E[\mathbf{a}/\mathbf{r}]$ is linear (this is true for a larger class of distribution). Frequently, $E[\mathbf{a}/\mathbf{r}]$ is assumed to be linear, which in general will result in a suboptimum estimator.

(2) Maximum a posteriori probability estimator (MAP)
 The estimator $\mathbf{a}(\mathbf{r})$ is selected so that $p(\mathbf{a}/\mathbf{r}) = \max p(\mathbf{r}/\mathbf{a})p(\mathbf{a})$. In this case, the risk function is assumed to be uniform (no weighting of errors).

Maximum likelihood estimation

In this model we assume that $p(\mathbf{r}/\mathbf{a})$ is known. This procedure is equivalent to MAP with uniform $p(\mathbf{a})$, that is, it is a Bayes procedure without assumptions (1) and (3).

Philosophical differences between MMSE and SA approaches

Consider the scalar case $\mathbf{a} = a$ and $\mathbf{r} = r$. The regression equation $\widehat{a}(r) = E[a/r]$ may be typical of the form shown in Fig. 8.1. In the SA, the same type of problem is approached from a different point of view. Consider noisy observables $T(r)$, representing an empirical version of a performance criterion $f(a)$, is to be minimized, maximized, or a root of it found (typically, see Fig. 8.2). The parameter a is to be estimated sequentially.

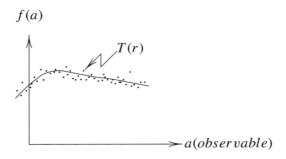

Figure 8.2. Stochastic approximation.

Comments:
- Sequential versions of the classical estimation techniques are available but are tractable only in very special cases, for example, when $p(r/a)$ and $p(a)$ are related in a reproducing manner. Sequential procedures offer computational simplification, superior performance, the possibility of intermediate decisions, and a variable dynamic observation time flexibility.
- Stochastic approximation is a distribution-free sequential estimation procedure; that is, there is no need for specific knowledge of $p(\mathbf{a})$ or $p(\mathbf{r}/\mathbf{a})$ (though general constraints on $p(\mathbf{r}/\mathbf{a})$ are required), and any partial knowledge can be of help in (1) placing the initial guess and (2) improving the speed of convergence.

Finally, one can relate the performance criteria for SA versus classical estimation as follows

Classical Estimation		SA	
Unbiased			
Consistent	$\Longleftrightarrow$	Convergence	(8.2)
Efficiency	$\Longleftrightarrow$	Speed of Convergence	
		(How fast $\sigma^2_{\text{error}} \longrightarrow 0$)	

8.2 Stochastic approximation procedures

Suppose a recursive sequence $\langle X_i \rangle$, $i = 1 \longrightarrow \infty$ is defined by choosing first X_1 and defining the successive X_i by

$$X_{n+1} = T_n(X_1, X_2, \ldots, X_n) \overset{\triangle}{=} T_n(X_n) \tag{8.3}$$

It is desired to choose the mapping sequence $\langle T_n \rangle$ so that $\lim_{n \to \infty} X_n = \theta$ where θ is the desired goal, for example, θ might be a root of an equation or the point at which a functional $f(\mathbf{r})$ assumes its minimum. We consider both the deterministic and stochastic cases.

8.2.1 Deterministic case

In the deterministic case, convergence is proven by the contraction mapping theorem.

> **Theorem 8.1** *If* $|T_n(x_n) - \theta| \leq F_n |x_n - \theta|$ *with* $\langle F_n \rangle$ *such that (a) there exists N such that $n \geq N \Longrightarrow 0 < F_n < 1$ and (b) $\sum_{n=N}^{\infty}(1 - F_n) = \infty$, then $\lim_{n \to \infty} X_n = \theta$.*

Here, condition (a) permits accelerated or other strategies for $1 \leq n < N$ while (b) permits $F_n \to 1$, for example, $F_n = (1 - \frac{1}{n})$.

> *Proof:* $|X_{n+1} - \theta| = |T_n(x_n) - \theta| \leq F_n |X_n - \theta| \leq F_n F_{n-1} |X_{n-1} - \theta| \leq \prod_{i=1}^{n} F_i |X_1 - \theta|$

Of course, we do not consider the trivial case where $\prod_{i=1}^{n} F_i \to 0$.

Thus, we seek the conditions that ensure $\lim_{n \to \infty}(\prod_{i=1}^{n} F_i) = 0$. Consider $n > N$, that is, $0 < F_n < 1$, we have

$$\prod_{i=N}^{n} F_i = \prod_{i=N}^{n}(1 - (1 - F_i)) \leq \prod_{i=N}^{n} e^{-(1-F_i)} = e^{-\sum_{i=N}^{n}(1-F_i)} \tag{8.4}$$

where we used the well-known equality $(1 - x) \leq e^{-x}$.

Hence, $\lim_{n \to \infty} \prod_{i=N}^{n} F_i = e^{-[\lim_{n \to \infty} \sum_{i=N}^{n}(1-F_i)]} = 0$, or $\lim_{n \to \infty}(\prod_{i=1}^{n} F_i) = 0 \Longrightarrow \lim_{n \to \infty} X_n = \theta$. QED.

The preceding theorem is not a specific scheme but a general statement of sufficient conditions to ensure convergence of a given iterative scheme to a sought value.

8.2.2 Stochastic case

We consider here the theorem due to Dvoretzky.[95] It provides a set of sufficient conditions to ensure convergence, in the mean square sense and with probability one, of an iterative scheme to a sought value.

We present Dvoretzky's theorem without proof. Later, we consider a simplified version that is useful in practice.

Let $\langle \alpha_n \rangle$, $\langle \beta_n \rangle$, and $\langle \gamma_n \rangle n = 1, 2, \ldots$ be nonnegative real sequences satisfying the following conditions

Assumption 1: Conditions for deterministic convergence.
(1) $\lim_{n \to \infty} \alpha_n = 0$: overshoot constraint sequence
(2) $\sum_{i=1}^{n} \beta_i < \infty$: initial strategy flexibility
(3) $\sum_{i=1}^{n} \gamma_i = \infty$: effort to correct undershoot
    Let X be a real number and T_n a measurable transformation such that
(4) $|T_n(X_1, X_2, \ldots, X_n) - \widehat{X}| \leq \max\{\alpha_n, [(1 + \beta_n)|X_n - \widehat{X}| - \gamma_n]\}$ for all real $X_1, X_2, \ldots, X_n$

Assumption 2: Conditions for noise dissipation.
Assume the following conditions with probability 1.
(1) $E[X_1^2] < \infty$: constraint on spread of initial guess
(2) $\sum_{i=1}^{n} E[r_i^2] < \infty$: ensures dissipation of noise perturbation
(3) $E[r_n] = 0$: ensures convergence to the right value

> **Theorem 8.2** *Consider the nonnegative real sequences $\langle \alpha_n \rangle$, $\langle \beta_n \rangle$, and $\langle \gamma_n \rangle$ satisfying assumptions 1 and 2, then*
> *(a) $\lim_{n \to \infty} E[(X_n - \widehat{X})^2] = 0$, that is, $X_n \to \widehat{X}$ in mean square sense*
> *(b) $Pr[\omega : \lim_{n \to \infty} X_n = \widehat{X}] = 1$, that is, $X_n \to \widehat{X}$ with probability 1*

The theorem is useful in applications where the iterative scheme can be represented in the form

$$\begin{cases} X_{n+1} = T_n(X_1, X_2, \ldots, X_n) + r_n \\ \\ r_n : \text{ noise component} \end{cases} \qquad \text{for } n > 0 \qquad (8.5)$$

Here, the sequences $\langle \alpha_n \rangle$, $\langle \beta_n \rangle$, and $\langle \gamma_n \rangle$ need not be independent of the observations $X_1, X_2, \ldots, X_n$ and can be replaced by $\alpha_n = \alpha_n(X_1, X_2, \ldots, X_n)$; $\beta_n = \beta_n(X_1, X_2, \ldots, X_n)$ and $\gamma_n = \gamma_n(X_1, X_2, \ldots, X_n)$ where again the constraints on these sequences are similar to (1), (2), and (3) [see Dvoretzky, pp. 40–41]. Also, bias in the noise component can be tolerated as long as it tends to vanish as the experimentation proceeds, that is, the theorem holds if condition (3) is replaced by $\sum_{i=1}^{n} E[r_i] = 0$.

We now consider a special case of the preceding theorem. The three sequences $\langle \alpha_n \rangle$, $\langle \beta_n \rangle$, and $\langle \gamma_n \rangle$ are now lumped together into a single sequence $\langle F_n \rangle$, which is the sequence in the contraction mapping theorem.

> **Theorem 8.3** *If (1) $\langle X_n(\omega) \rangle$, $n = 1 \to \infty$, and $\langle Y_n(\omega) \rangle$, $n = 1 \to \infty$ denotes sequences of r.v defined over $(\Omega, \mathcal{F}, P)$, (2) $T_n[X_1(\omega), \ldots, X_n(\omega)]$ will be a measurable transformation of its arguments. (3) $\langle X_n \rangle$ and $\langle Y_n \rangle$*

are defined by $X_{n+1}(\omega) = T_n[X_1(\omega), \ldots, X_n(\omega)] + Y_n(\omega)$. *(4) There exists a sequence* $\langle F_n \rangle$ *of positive real numbers such that* $\prod_{n=1}^{\infty} F_n = 0$ *and* $\prod_{n=r}^{s} F_n < K < \infty$ *for all* $s > r$ *and* $|T_n(X_1, X_2, \ldots, X_n) - \theta| \le F_n|X_n - \theta|$. *(5)* $E[X_1^2] < \infty$ *and* $\sum_{n=1}^{\infty} E[Y_n^2] < \infty$ *and (6)* $E[Y_n/X_1, X_2, \ldots, X_n] = 0$ *for all* n, *then*

$$\lim E[(X_n - \theta)^2] = 0, \text{ that is, } \lim_{n \to \infty} X_n = \theta.$$

and

$$P[\omega : \lim_{n \to \infty} X_n(\omega) = 0] = 1, \text{ that is, } X_n \to \theta \text{ with prob 1}$$

(see reference [86]).

Here, condition (4) is a fixed point condition for deterministic component convergence and condition (5) ensures random error dissipation, while (6) assumes that the observations are unbiased.

Proof: Mean square sense convergence. Let

$$V_n^2 = E[(X_n - \theta)^2] \text{ and } \sigma_n^2 = E[Y_n^2] \tag{8.6}$$

$$V_{n+1}^2 = E[(T_n(X_1, X_2, \ldots, X_n) + Y_n - \theta)^2]$$
$$= E[(T_n - \theta)^2] + E[Y_n^2] \tag{8.7}$$

Using constraints (4) and (6), Eq. (8.7) can be written as

$$V_{n+1}^2 \le F_n^2 V_n^2 + \sigma_n^2 \tag{8.8}$$

Iterating, we obtain

$$V_{n+1}^2 \le F_n^2 V_n^2 + \sigma_n^2 \le F_n^2(F_{n-1}^2 V_{n-1}^2 + \sigma_{n-1}^2) + \sigma_n^2$$

$$= F_n^2 F_{n-1}^2 F_{n-2}^2 \cdots F_1^2 V_1^2 + F_n^2 F_{n-1}^2 \cdots F_2^2 \sigma_1^2$$

$$+ \ldots + F_n^2 F_{n-1}^2 \sigma_{n-2}^2 + F_n^2 \sigma_{n-1}^2 + \sigma_n^2$$

$$= (\sigma_n^2 + \sigma_{n-1}^2 F_n^2 + \ldots + \sigma_m^2 F_{m+1}^2 \cdots F_n^2) + \sigma_{m-1}^2 F_m^2 \cdots F_n^2$$

$$+ \ldots + \sigma_1^2 F_2^2 F_3^2 \cdots F_n^2 + V_1^2 F_1^2 F_2^2 \cdots F_n^2$$

$$\le \left(\sum_{j=m}^{n} \sigma_j^2 \right) \left(\max_{m \le k \le n} \prod_{j=k+1}^{n} F_j^2 \right) + \left(V_1^2 + \sum_{j=1}^{m-1} \sigma_j^2 \right) \left(\max_{1 \le k \le m} \prod_{j=k}^{m} F_j^2 \right)$$

$$\le K \sum_{j=m}^{\infty} \sigma_j^2 + \left(V_1^2 + \sum_{j=1}^{\infty} \sigma_j^2 \right) \left(\max_{1 \le k \le m} \prod_{j=k}^{n} F_j^2 \right) \tag{8.9}$$

whereby using Eq. (8.5) for all $\epsilon > 0$, there exists m such that $K \sum_{j=m}^{\infty} \sigma_j^2 < \frac{\epsilon}{2}$ and $(V_1^2 + \sum_{j=1}^{\infty} \sigma_j^2) < \infty$. For m fixed, there exists N such that for every $n \geq N \max_{1 \leq k \leq m} \prod_{j=k}^{n} F_j^2 \leq (\epsilon/2(V_1^2 + \sum_{j=1}^{\infty} \sigma_j^2))$; that is, for any $\epsilon > 0$, there exists a N such that for every $n > N$, $V_n < \epsilon$. Therefore, $\lim_{n \to \infty} V_n = 0$, that is, $\lim_{n \to \infty} X_n = \theta$.

The proof suggests the following strategy in proving the convergence of a specific scheme

(1) Represent the scheme as in Eq. (8.3).
(2) Test the random part.
(3) Test the deterministic part.

Next, we consider two important stochastic procedures.

8.2.3 The scalar version of the Robbins–Monro procedure

The present recursive procedure was proposed by Robbins and Monro[93] for estimating the root θ of the equation $M(x) = \alpha$, where α is a given number.

We assume that instead of measuring $M(x)$, we actually observe $Z(x) = M(x) + N(x)$ where $N(x)$ is the additive noise term, independently distributed from sample to sample. In addition, we assume $E[Z(x)] = M(x)$. The recursive scheme is expressed as follows

$$X_{n+1} = X_n - a_n Z_{X_n} \tag{8.10}$$

where $\langle a_n \rangle$ is a sequence of positive numbers.

Theorem 8.4 *If (a)* $\sigma_{Z_x}^2 \leq \sigma^2 < \infty$ *for all* x, *(b)* $0 < A \leq \frac{M(x)}{X - \theta} \leq B < \infty$, *(c)* $\sum_{i=1}^{\infty} a_i = \infty$, *(d)* $\sum_{i=1}^{\infty} a_i^2 < \infty$ *and (e)* $E[X_1^2] < \infty$, *then* $X_n \to \theta$ *in mean square sense and wp1 (see reference [86]).*

Proof: We follow the strategy outlined in the previous section.
(I) *Partitioning*

$$X_{n+1} = X_n - a_n Z_{x_n} = [X_n - a_n M(X_n)] + a_n [M(X_n) - Z_{X_n}] \tag{8.11}$$

Following the model in Theorem 8.3, we have

$$T_n(X_n) = X_n - a_n M(X_n) \tag{8.12}$$

and

$$Y_n = a_n [M(X_n) - Z_{x_n}] \tag{8.13}$$

(II) *Random part*

Using (a) and (d), we have

$$\sum_{i=1}^{\infty} E[Y_i^2] = \sum_{i=1}^{\infty} a_i E[Y_i(M(X_i) - Z_{X_i})] = -\sum_{i=1}^{\infty} a_i E[Y_i Z_{X_i}]$$

$$= -\sum_{i=1}^{\infty} a_i^2 E[M(x_i)Z(X_i)] + \sum_{i=1}^{\infty} a_i^2 E[Z_{X_i}^2]$$

$$= \sum_{i=1}^{\infty} a_i^2 E[Z_{X_i}^2] - \sum_{i=1}^{\infty} a_i^2 M^2(X_i)$$

$$\leq \sigma^2 \sum_{i=1}^{\infty} a_i^2 < \infty. \tag{8.14}$$

Thus, $\sum_{i=1}^{\infty} a_i^2 < \infty$ is needed to ensure that the noise is dissipated.

(III) *Deterministic part*

$$|T_n(X_n) - \theta| = |X_n - a_n M(X_n) - \theta| \leq |X_n - \theta| \sup_X \left| 1 - a_n \frac{M(x)}{X - \theta} \right|$$

$$\leq |X_n - \theta| \max(1 - a_n A, a_n B - 1) \tag{8.15}$$

by using assumption (b).

Since $a_n \to 0$ and $0 < A < B$, there exists a N such that $0 < F_n = (1 - a_n A)$. It suffices now to consider $n > N$ when $|T_n(X_n) - \theta| \leq (1 - a_n A)|X_n - \theta| = F_n|X_n - \theta|$. Also, $\sum_n a_n = \infty \to \prod_n F_n = 0$. Therefore, $X_n \to \theta$ wp1 and m.s.

8.2.4 The Kiefer–Wolfowitz procedure

The present procedure seeks the location θ of the maximum (or minimum) of $M(x)$. With the observable $Z(x) = M(x) + N(x)$, the recursive scheme is of the form

$$X_{n+1} = X_n - \frac{b_n}{c_n}[Z_{X_n+c_n} - Z_{X_n-c_n}] \tag{8.16}$$

where $\{b_n\}$ and $\{c_n\}$ are positive sequences. We assume that θ is the minimum sought of the regression function $M(x)$.

Theorem 8.5 *If (a1)* $\sigma_{Z_x}^2 \leq \sigma^2 < \infty$ *for all* x, *(a2)* $Z(X_n + c_n)$ *and* $Z(X_n - c_n)$ *are statistically independent, (b)* $\sum_{i=1}^{\infty} (\frac{b_n}{c_n})^2 < \infty$, *(c)* $\sum_{i=1}^{\infty} b_i = \infty$, *(d)* $\lim_{n \to \infty} c_n = 0$. *(e)* $0 < A \leq \left| \frac{M(X_n+c_n) - M(X_n-c_n)}{c_n(X_n-\theta)} \right| < B$, *and*

(f) $E[X_1^2] < \infty$, then

$$X_n \to \theta \text{ in m.s sense and wp1 (see reference [86])} \tag{8.17}$$

Here, (a1), (a2), and (b) are conditions for noise dissipation while (c), (d), and (e) are conditions for deterministic convergence.

Proof: Again, we follow the same partitioning strategy.
(I) *Partitioning*

$$X_{n+1} = X_n - \frac{b_n}{c_n}[Z(X_n + c_n) - Z(X_n - c_n)]$$

$$= \left[X_n - \frac{b_n}{c_n}\right][M(X_n + c_n) - M(X_n - c_n)] + \frac{b_n}{c_n}[M(X_n + c_n)$$

$$- M(X_n - c_n) + Z(X_n - c_n) - Z(X_n + c_n)] \tag{8.18}$$

where

$$T_n(x) = X_n - \frac{b_n}{c_n}[M(X_n + c_n) - M(X_n - c_n)] \tag{8.19}$$

and

$$Y_n = \frac{b_n}{c_n}[M(X_n + c_n) - M(X_n - c_n) + Z(X_n - c_n) - Z(X_n + c_n)] \tag{8.20}$$

(II) *Random part*

$$\sum_{i=1}^{\infty} E[Y_n^2] = \sum_{i=1}^{\infty} \frac{b_n^2}{c_n^2}(\sigma_{X_n+c_n}^2 + \sigma_{X_n-c_n}^2 - E[(Z(X_n + c_n)$$

$$- M(X_n + c_n))(Z(X_n - c_n) - M(X_n - c_n))]) \tag{8.21}$$

The last term on the right side is zero by assumption (a2) of Theorem 8.5. This implies that

$$\sum_{i=1}^{\infty} E[Y_n^2] = 2\sigma^2 \sum_{i=1}^{\infty} \frac{b_n^2}{c_n^2} < \infty \tag{8.22}$$

using (a1) and (b).
(III) *Deterministic part*

$$|T_n(X_n) - \theta| = \left|X_n - \frac{b_n}{c_n}[M(X_n + c_n) - M(X_n - c_n)] - \theta\right|$$

$$= |X_n - \theta|\left|1 - \frac{b_n}{c_n}\frac{(M(X_n + c_n) - M(X_n - c_n))}{(X_n - \theta)}\right|$$

$$\leq [max(1 - b_n A, b_n B - 1)]|X_n - \theta| \tag{8.23}$$

Let

$$F_n = [max(1 - b_n A, b_n B - 1)] \tag{8.24}$$

As before, it suffices now that $\sum_{i=1}^{\infty} b_n = \infty$. Therefore, converges $X_n \to \theta$ wp1 and in the mean square sense.

8.3 Stochastic approximation in least square estimation

Until now, the discussion has centered around the contraction mapping theorem in the deterministic case and Dvoretzky's theorem in the stochastic case. Specific procedures that were presented are the Robbins–Monro and Kiefer–Wolfowitz techniques for estimating the roots and location of the maximum (minimum) of a function, respectively. In the present context of linear models in image processing applications, we generally have to estimate the parameters in the model based on the least squares approach.

An interesting theorem proposed by Gladyshev[81] extended the SA techniques to least squares. Before we discuss the scalar case, we first present Sacks' theorem.[92]

> **Theorem 8.6** *Let* $\{a_n\}$ *be a sequence of positive numbers such that* $\sum_1^{\infty} a_n = \infty$, $\sum_1^{\infty} a_n^2 < \infty$. *Let* X_i *be a fixed number of arbitrary random varaibles such* $E[X_1^2] < \infty$ *and the sequence* $\{X_n\}$ *defined by the recursion*
>
> $$X_{n+1} = X_n - a_n(y(X_n) - \alpha) \tag{8.25}$$
>
> *where* $y(X_n)$ *is a random variable whose conditional distribution given* $x_1, x_2, \ldots, x_n$ *is the same as the distribution of* $y(x_n)$. *Let* $y(x) = M(x) + Z(x)$ *where the regression function* $M(x) = E[y(x)] = \alpha$ *has a unique solution for each real value* α. *Under the following assumptions:*
> (1) *M is a Borel measurable function and* $(x - \theta)(M(x) - \alpha) > 0$ *for all* $x \neq \theta$.
> (2) *For positive* k *and* k_1 *and for all* x,
> $k|x - \theta| \leq |M(x) - \alpha| \leq k_1|x - \theta|$.
> (3) *For all* x,
> $M(x) = \alpha + \alpha_1(x - \theta) + \delta(x, \theta)$ *where* $\delta(x, \theta) = o(|x - \theta|)$ *as* $x - \theta \to 0$ *and* $\alpha_1 > 0$.
> (4) $\sup_x E[Z^2(x)] < \infty$ *and* $\lim_{x \to \theta} E[Z^2(x)] = \sigma^2$.
> (5) $\lim_{R \to \infty} \lim_{\epsilon \to 0^+} \sup_{|x - \theta| < \{\epsilon|Z(x)| > R\}} \int Z^2(x)dt = 0$.
> *Let the sequence* $\langle a_n \rangle$ *such that* $a_n = A/n$ *for* $n > 0$ *and* A *is such that* $2kA > 1$. *Then,* $\sqrt{n}(X_n - \theta)$ *is asymptotically normally distributed with mean zero and variance* $A^2\sigma^2(2A\alpha_1 - 1)^{-1}$.

Assumptions 1–3 ensure convergence of the recursion without oscillation.

Proof: See reference [86].

8.3.1 Scalar case

Now, we are in a position to present Gladyshev's theorem in the scalar case.

Theorem 8.7 *Let $(\xi, v), (\xi_1, v_1) \dots$ be a sequence of two-dimensional real random vectors with $E[\xi^4] < \infty$ and $E[v^4] < \infty$. Further, let $X_1, X_2, \dots$ be a random sequence in which X_1 is an arbitrary real random variable and $X_2, X_3, \dots$ are determined using the iterative scheme*

$$X_{n+1} = X_n - (A/n)(\xi_n X_n - v_n)\xi_n \tag{8.26}$$

where $A(.) > 0$. Then, the sequence X_n converges wp1 to a value θ that minimizes the expression $E[(\xi\theta - v)^2]$. If in addition $AE[\xi^2] > \frac{1}{2}$ then the sequence $\sqrt{n}(X_n - \theta)$ is asymptotically normal with mean zero and variance

$$A^2(2AE[\xi^2] - 1)^{-1}E[\theta\xi^2 - \theta E[\xi^2] - \xi v + E[\xi v]] \tag{8.27}$$

Proof: The regression function is

$$M(X_n, \theta) = E[Y(X_n, \theta)] = E[(\xi_n X_n - v_n)\xi_n] \tag{8.28}$$

With $X_n \to \theta$, $\xi_n \to \xi$, and $v_n \to v$, one can use the similarity with the linear prediction approach to prove the first part of the theorem. Toward that end, consider $\Delta = \xi X_n - v$ as the error while ξ represents the data. Therefore, by virtue of the orthogonality of ξ and Δ, it implies that θ minimizes $E[(\xi\theta - v)^2]$. To prove that $X_n \to \theta$ wp1, it is sufficient to recall the proof of the scalar RM procedure. To prove the second part of the theorem, we make use of Sacks' theorem with the assumption $\alpha = 0$. We therefore have to verify that the assumptions of the theorem are applicable. Let $y(X_n) = \xi_n^2 X_n - \xi_n v_n$. The regression function is given by

$$M(X_n) = E[y(X_n)] = E[\xi_n^2]X_n - E[\xi_n v_n] \tag{8.29}$$

and the noise term $Z(X_n)$ is therefore equal to

$$Z(X_n) = y(X_n) - M(X_n) = \xi_n^2 - X_n v_n - E[\xi_n^2]X_n + E[\xi_n v_n] \tag{8.30}$$

which implies $E[Z(x)] = 0$ for all x.

Consider the variable $v_n = \theta\xi_n + e_n$ where e_n are i.i.d zero mean symmetric random variables. Substituting in Eq. (8.29), we obtain

$$M(X_n) = E[\xi_n^2](X_n - \theta) \tag{8.31}$$

which shows the explicit dependence between $M(X_n)$ and θ. In addition, Eq. (8.31) implies that $M(X_n) = 0$ has a unique root at $x = \theta$. Moreover, because $M(X_n)$ is linear, assumptions 1–3 are satisfied. Since we assume $E[\xi^4] < \infty$ and $E[v^4] < \infty$, it follows that assumption 4 is satisfied.

Let $\rho^2 = E[(\xi_n^2 - \theta v_n - E[\xi_n^2]\theta + E[\xi_n v_n])^2]$. Since the assumptions of Sacks' theorem are satisfied, it implies that the sequence $\sqrt{n}(X_n - \theta)$ is asymptotically normal with the variance obtained by substituting α_1 for $E[\xi^2]$ and ρ^2 for σ^2 in the variance in Sacks' theorem.

8.3.2 Vector case

We now extend the discussion to include the vector case by assuming that multiple observations $\mathbf{y}_k$ of $\mathbf{y}$ are available. Denoting k, $k = 1, 2, \ldots, K$ the order of replication, the vector form of stochastic approximation minimum variance least squares SAMVLS algorithm is given by

$$\widehat{\beta}_{k+1} = \widehat{\beta}_k - \frac{A_k}{k} \mathbf{X}^T [\mathbf{X}\widehat{\beta}_k - \mathbf{y}_k] \quad k = 1, 2, \ldots, K \tag{8.32}$$

where the data model is as follows

$$\mathbf{y}_k = \mathbf{X}^T \beta_k + \mathbf{V}_k \tag{8.33}$$

By comparing the present formulation to that of the scalar case, ξ becomes the design matrix $\mathbf{X}$ while θ is now a vector β of r parameters to be estimated. In Eq. (8.32), A_k is an $(r \times r)$ diagonal matrix whose elements are to be estimated.

Let $Y(\widehat{\beta}, \beta) = \mathbf{X}^T(\mathbf{X}\widehat{\beta} - \mathbf{y}_k)$. The regression function is then

$$M(\widehat{\beta}, \beta) = \mathbf{X}^T \mathbf{X}(\widehat{\beta} - \beta) \tag{8.34}$$

which implies that $M(\widehat{\beta}, \beta) = 0$ has a unique solution at $\beta = \widehat{\beta}$.

Theorem 8.8 *Let $r_1 > r_2 > \ldots > r_r > 0$ be eingenvalues of* $\mathbf{AB}$; *where* $\mathbf{B} = \mathbf{X}^T\mathbf{X} = \mathbf{I}_{r \times r}$. *Then, $\sqrt{k}(\widehat{\beta}_k - \beta)$ is asymptotically normal with mean zero and covariance matrix $\mathbf{Q}$, where $\mathbf{Q}$ is a diagonal matrix whose elements are $a_{ii}^2 \sigma_{ii}^2 [2a_{ii} - 1]^{-1}$.*

Proof: See reference [73].

8.3.3 Small sample theory

Consider the linear model

$$\mathbf{y}_k = \mathbf{X}^T \beta + \mathbf{V}_k \tag{8.35}$$

The parameters of the model are recursively estimated using the following scheme

$$\widehat{\beta}_{k+1} = \widehat{\beta}_k - \frac{A(.)}{k}\mathbf{X}^T[\mathbf{X}\widehat{\beta}_k - \mathbf{y}_k] \quad k = 1, 2, \ldots, K \tag{8.36}$$

The recursion is stopped after a finite number of iterations K. One method for the selection of the required number of iterations K is to compute simultaneously or before the SAMVLS is initiated, the LS estimates of the parameters. The error term $\beta_{SAMVLS} - \beta_{LS}$ is compared to an acceptable error bound. The recursion is terminated when the desired accuracy is attained.

However, the resulting estimate may exhibit a small bias depending on the selected value of K. This should not dramatically affect the test of hypotheses. When the noise distribution is well behaved, Kadar and Kurz[12] recommend using SAMVLS as a channel sampler in a reference channel configuration to allow a learning process for unusual noise disturbances.

Small sample effect in the scalar case

In practice, what is available is a small sample size. Therefore, we need to consider the size effect on the behavior of the estimate θ. The observations are in the form

$$Y(X_n, \theta) = M(X_n, \theta) + Z(X_n) \tag{8.37}$$

where $Y(X_n, \theta) = (\xi_n X_n - \nu_n)\xi_n$. The regression function is given by

$$M(X_n, \theta) = E[Y(X_n, \theta)] = E[\xi_n^2]X_n - E[\xi_n \nu_n] \tag{8.38}$$

Let

(1) $\alpha_{1n} = E[\xi_n^2]$
(2) $b_n = -E[\xi_n \nu_n]$
(3) $\nu_n = \theta \xi_n + V_n$ where V_n are i.i.d. zero mean symmetric random variables.

Hence, $M(X_n, \theta) = \alpha_{1n}(X_n - \theta) + b_n$. The noise term is

$$Z(X_n) = \xi_n^2 X_n - \xi_n \nu_n - E[\xi_n^2]X_n + b_n \tag{8.39}$$

The root θ of $M(x)$ is estimated via the following recursion

$$X_{n+1} = X_n - a_n[\alpha_{1n}(X_n - \theta) + \epsilon_n \rho] \tag{8.40}$$

where ϵ_n is sequence of i.i.d. random variables satisfying $E[\epsilon_n] = 0$ and $E[\epsilon_n^2] = 0$
Therefore, $Z(X_n) = \epsilon_n \rho$ with mean $E[Z(X_n)] = 0$ and $E[z^2(x_n)] = \rho^2$ as $X_n \to \theta$. Substituting $\delta_n = X_n - \theta$, we can rewrite Eq. (8.40) as

$$\delta_{n+1} = \delta_n \left(1 - \frac{a}{n}\right) - \left(\frac{a}{n}\right)\left(\frac{\rho}{\alpha_{1n}}\right)\epsilon_n \tag{8.41}$$

where we assume constant gain coefficients with $a_n = a/(n\alpha_{1n})$.

By iterating, it can be shown that

$$\delta_n = \delta_1 \prod_{k=1}^{n-1} \left(1 - \frac{a}{k}\right) - \rho a \sum_{i=1}^{n-1} \frac{\epsilon_i}{i\alpha_{1i}} \prod_{k=1+i}^{n-1} \left(1 - \frac{a}{k}\right) \tag{8.42}$$

The expected bias at step n of the iteration is then

$$E[\delta_n] = E[X_n - \theta] = \delta_1 \prod_{k=1}^{n-1} \left(1 - \frac{a}{k}\right) \tag{8.43}$$

In addition, the variance of δ_n is given by

$$\text{Var}(\delta_n) = \rho^2 \sum_{i=1}^{n-1} \left[\frac{a}{i} \prod_{k=i+1}^{n-1} \left(1 - \frac{a}{k}\right) \right] \frac{1}{\alpha_{1i}^2} \tag{8.44}$$

Using the approximation $1 - a/k \simeq e^{-ak}$ and letting $a_n = A/n$ where $A = a/E[\xi^2]$, one can show that

$$\lim_{n \to \infty} E[\delta_n] = 0 \tag{8.45}$$

and

$$\lim_{n \to \infty} \text{Var}[\delta_n] = \frac{\rho^2 A^2}{n\left[2AE[\rho^2] - 1\right]} \tag{8.46}$$

with

$$AE[\rho^2] > 1/2 \tag{8.47}$$

Consequently, the expected squared error of X_n is of order $0(1/n)$. Sacks[92] shows that $\sqrt{n}(X_n - \theta)$ is asymptotically normal. By minimizing Eq. (8.46), a suitable choice for A is

$$A = \frac{1}{E[\rho^2]} \tag{8.48}$$

and the corresponding minimum of the variance in Eq. (8.46) is

$$\lim_{n \to \infty} \text{Var}(\delta_n) = \frac{\rho^2}{nE[\rho^2]} \tag{8.49}$$

8.4 Robust recursive estimation

The F-test is robust with respect to the departure from the normality assumption of the noise component.[2] This is true in particular for a large class of imaging applications discussed in this book. The lack of robustness in the face of correlated noise environment has been dealt with in previous chapters by the introduction of an orthogonal transformation and is not discussed here.

In the presence of outliers the parameter estimates are generally biased, which in turn affects the test of hypotheses. The techniques of SAMVLS of Section 8.3 are still applicable by using their robustized versions. This guarantees that the estimators are asymptotically normal, in which case the F-test can be used without further modification.

A large body of literature exists on robust approaches to least squares estimation. In this section we discuss only two approaches: Huber's M-estimators and the rank statistic preprocessor.

8.4.1 Rank statistic preprocessor

Let $X_1, X_2, \ldots, X_m$ and $Y_1, Y_2, \ldots, Y_n$ be ordered observations from continuous, not necessarily symmetric, cumulative distribution functions (CDFs) $F(x)$ and $G(x)$. A class of rank tests is based on linear combinations of the observed vector of observations $\mathbf{X}$ and $\mathbf{Y}$. The discussion in this section centers around the one-sample and two-sample tests.

Basic definitions

Let the hypothesis and alternative H and K be defined as follows

$$
\begin{aligned}
H : & \quad F(x) = G(x) \\
K : & \quad F(x) \neq G(x)
\end{aligned}
\tag{8.50}
$$

The nonparametric statistics of the Chernoff–Savage class[91] are of the form

$$
T_N = \frac{1}{mn} \sum_{i=1}^{N} a_i Z_i
\tag{8.51}
$$

where $N = m + n$.

The coefficients a_i are called "scores" and are selected based on certain optimality criteria.

One-sample case

Consider the ordered set $\{\mathbf{X}\}$ and let $N = m$. The class of one-sample linear ranked statistic is of the same form as the Chernoff–Savage class. The indicator variables Z_i are defined as follows

$$
Z_i = \begin{cases} +1 & \text{if } X_i < 0 \\ 0 & \text{otherwise} \end{cases}
\tag{8.52}
$$

where the ordered observations are assumed to be from a population with symmetric CDF $F(x)$. Two statistics are of interest under the one-sample theory: the Wilcoxon statistic and the Wilcoxon symmetric statistic (WSNRS). [88–91]

Wilcoxon statistic

Let k_i be the rank of $|X_i|$ in the ranked set of observations. The Wilcoxon statistic is defined as follows

$$S = \frac{1}{m} \sum_{i=1}^{m} k_i \, sgn(X_i) \tag{8.53}$$

Another form of Eq. (8.53) often used in practical implementation is

$$S^* = \sum_{i=1}^{m} \sum_{j=1}^{i-1} U(X_i + X_j) + \sum_{j=1}^{m} U(X_j) \tag{8.54}$$

Wilcoxon symmetric statistic

Using the relation $sgn(z) = 2U(z) - 1$ and substituting in Eq. (8.54), we obtain the equivalent statistic

$$S = \sum_{i>j}^{m} sgn(X_i + X_j) + \sum_{j=1}^{m} sgn(X_j) \tag{8.55}$$

where under H, $E[S] = 0$ and $\mathrm{Var}[S] = m(m+1)(2m+1)/6$.
Under H and K,

$$\lim_{m \to \infty} P\left[\frac{S - E[S]}{\sqrt{\mathrm{Var}\,S}} \leq u \right] = \Phi(u) \tag{8.56}$$

where $\Phi(u)$ is the unit normal CDF. S reaches asymptotic normality with about $m = 20$ samples.[88]

Two-sample case

The vector of observations $\mathbf{Y}$ is obtained under no signal conditions during some learning intervals, while $\mathbf{X}$ is obtained during the decision interval. In the two-sample problem, the indicator variables are defined as follows

$$Z_i = \begin{cases} +1 & \text{if } i^{\text{th}} \text{ smallest sample of } \{\mathbf{X} \cup \mathbf{Y}\} \in \{\mathbf{X}\} \\ 0 & \text{otherwise} \end{cases} \tag{8.57}$$

By replacing a_i with $i/(N+1)$, we obtain the usual U-form

$$U = \frac{1}{mn} \sum_{i=1}^{m} \sum_{j=1}^{n} U(X_i - Y_j) \tag{8.58}$$

Under H, $E[U] = mn/2$ and $\mathrm{Var}[U] = mn(m+n+1)/12$.

The Mann–Whitney–Wilcoxon statistic (MWWNS) is the symmetric version of the U-form and is directly obtained from Eq. (8.58)

$$W = \frac{1}{mn} \sum_{i=1}^{m} \sum_{j=1}^{n} \mathrm{sgn}(X_i - Y_j) \tag{8.59}$$

which under H has $E[W] = 0$ and $\mathrm{Var}[W] = mn(m+n+1)/3$. In addition, under H and K

$$\lim_{m \to \infty} P\left[\frac{W - E[W]}{\sqrt{\mathrm{Var}\,W}} \le u \right] = \Phi(u) \tag{8.60}$$

Asymptotic normality is reached for values of m and n as low as 8,[88] which is far less than the size required for the one-sample class of tests.

8.4.2 Huber's M-Estimator

Let $X_1, X_2, \ldots, X_n$ be i.i.d random variables with CDF $F(x - \theta)$[1] where F is symmteric.[2] Huber[85] proposed a maximum likelihood type estimator of θ, called M-estimator, which satisfies the equation

$$\sum_{i=1}^{n} \psi(X_i - M) = 0 \tag{8.61}$$

and $\psi(x)$ is such that $\psi(-x) = -\psi(x)$. If in addition $\psi(x)$ is monotonic, the estimate is uniquely determined. The estimator is asymptotically normal with mean θ, and the asymptotic variance is such that

$$\mathrm{Var}(n^{1/2} M) = V_M(\psi, F) = \frac{\int \psi^2(x) f(x) dx}{\left[\int \psi'(x) f(x) dx \right]^2}$$

$$= \frac{E_F[\psi^2]}{\left[E_F[\psi'] \right]^2} \ge \frac{1}{\int (f'(x)/f(x))^2 f(x) dx}$$

Huber shows there exists for $V_M(\psi, F)$ a saddle point for ψ_0 and the class C of distributions of the form $F = (1 - \epsilon)G + \epsilon H$, where ϵ is such that $0 \le \epsilon \le 1$ is a

---

[1] We consider the location estimate problem in the present section.
[2] That F is symmetric allows the use of a skew-symmteric function $\psi(x)$.

fixed number, G is fixed and symmetric, and H is a variable CDF. For the special case of ϵ-contaminated normal class, there exists a ψ_0 for a distribution F_0 that defines an M-estimator that is asymptotically normal. In addition, for all F in the class of distributions, there exists a saddle point $V_M(\psi_0, F_0)$ satisfying

$$V_M(\psi, F_0) \leq V_M(\psi_0, F_0) \leq V_M(\psi_0, F) \tag{8.62}$$

where $V_M(\psi_0, F_0) = I(F_0)$ and $I(F)$, the Fisher information, is such that

$$I(F) = \int (f'(x)/f(x))^2 f(x)dx < \infty \text{ and } I(F_0) < I(F) \tag{8.63}$$

8.4.3 Robust SAMVLS

We now consider the question of fitting the above robust procedures to the theory of SAMVLS. We begin with the one-sample case. Then, we extend the analysis to the M-estimator (LLIF).[3] Finally, in the next section, we illustrate the theory by deriving the parameter estimates for the two-way model using the WSNRS statistic.

One-sample case

The SAMVLS recursion in the robust case is of the form

$$\widehat{\beta}_{k+1} = \widehat{\beta}_k - \frac{1}{k} A_k S^m(Y(\widehat{\beta}_k, \beta)) \tag{8.64}$$

where $S^m(.)$ is the symmetric Wilcoxon statistic and

$$Y(\widehat{\beta}_{k+i}, \beta) = \mathbf{X}^T[\mathbf{X}(\widehat{\beta}_{k+i} - \beta) + V_{k+i}] \ i = 1, 2, \ldots, m. \tag{8.65}$$

The class of noise distributions is assumed to be symmetric.

Huber's M-estimator

The LLIF robustized SAMVLS recursion is of the form

$$\widehat{\beta}_{k+1} = \widehat{\beta}_k - \frac{A_L}{k} l_k \left[Y\left(\widehat{\beta}_k, \beta\right) \right] \tag{8.66}$$

where

$$l_k(z) = \begin{cases} z & |z| < k \\ k \ sgn(z) & |z| \geq k \end{cases} \tag{8.67}$$

For each of the above cases, one has to prove convergence of the recursion. Also, it is necessary to evaluate the optimum gain coefficients, which is accomplished by

---

[3]Light limiter influence function.

first deriving the regression function and then taking the derivative at the root. We illustrate the procedure in the next section.

8.5 An illustrative example

In this section we present a derivation of the parameters estimates for a reparameterized, replicated, full-rank two-way design. We assume the order of replication is K and the effects are fixed. Hence, the approach is based on the vector form of Gladyshev's theorem. In addition, we assume that the mask is of size 2×2, which implies that the vector of effects β is a (3×1) vector. The model is thus of the form

$$\mathbf{y} = \mathbf{X}^T \beta + \mathbf{V} \tag{8.68}$$

where $\mathbf{X}$ is a (3×4) matrix.

$\mathbf{V}$ is a (4×1) additive symmetrically distributed noise vector, i.i.d. for each replication k, $k = 1, 2, \ldots, K$. The recursion can now be written as

$$\widehat{\beta}_{k+1} = \widehat{\beta}_k - \frac{A_k}{k} \mathbf{S}^m \mathbf{Y}(\widehat{\beta}_k, \beta) \tag{8.69}$$

where $S_l^m(.)$ for each vector component l, $l = 1, 2, 3$ is a symmetric version of WSRNS. The noise distributions are by the assumption restricted to symmetric. In addition, $\mathbf{Y}(\widehat{\beta}_k, \beta)$ in Eq. (8.69) is given by

$$\mathbf{Y}(\widehat{\beta}_{k+i}, \beta) = \mathbf{X}^T \left[\mathbf{X}(\widehat{\beta}_{k+i} - \beta) + V_{k+i} \right] \qquad i = 1, 2, \ldots, m. \tag{8.70}$$

At this stage, it remains to define the optimum gain coefficients A_k. As mentioned before, this is accomplished by first deriving the robust form of the regression function and then taking the derivative at $\widehat{\beta} - \beta$. Toward that end, define

$$U_l^m = \sum_{i=1}^m \sum_{j=1}^{i-1} U(Z_i + Z_j) + \sum_{i=1}^m U(Z_i) \qquad l = 1, 2, 3. \tag{8.71}$$

which are related to the components $S_l^m()$ of $\mathbf{S}^m$ by

$$\begin{pmatrix} S_1^m(.) \\ S_2^m(.) \\ S_3^m(.) \end{pmatrix} = \begin{pmatrix} 2U_1^m(.) - m(m+1)/2 \\ 2U_2^m(.) - m(m+1)/2 \\ 2U_3^m(.) - m(m+1)/2 \end{pmatrix} \tag{8.72}$$

It can be shown by extending the results obtained in the scalar case[80] that

$$
E(U^m) = \begin{pmatrix} \sum\limits_{i=1}^{m} \sum\limits_{j=1}^{i-1} \int\limits_{-4(\beta_1-\widehat{\beta}_{1k})}^{\infty} f_1(x)dx + \sum\limits_{i=1}^{m} \int\limits_{-4(\beta_1-\widehat{\beta}_{1k})}^{\infty} g_1(x)dx \\[2em] \sum\limits_{i=1}^{m} \sum\limits_{j=1}^{i-1} \int\limits_{-4(\beta_2-\widehat{\beta}_{2k})}^{\infty} f_2(x)dx + \sum\limits_{i=1}^{m} \int\limits_{-4(\beta_2-\widehat{\beta}_{2k})}^{\infty} g_2(x)dx \\[2em] \sum\limits_{i=1}^{m} \sum\limits_{j=1}^{i-1} \int\limits_{-4(\beta_3-\widehat{\beta}_{3k})}^{\infty} f_3(x)dx + \sum\limits_{i=1}^{m} \int\limits_{-4(\beta_3-\widehat{\beta}_{3k})}^{\infty} g_3(x)dx \end{pmatrix} \quad (8.73)
$$

where $f_l(x)$, $l = 1, 2, 3$ are the components of the pdf of $\mathbf{X}^T(\mathbf{V}_i + \mathbf{V}_j)$ and $g_l(x)$, $l = 1, 2, 3$ are the components of the pdf of $\mathbf{X}^T \mathbf{V}_i$, respectively. Hence, the slope of the regression function is given by

$$
\mathbf{B} = \begin{pmatrix} \alpha_{11} & 0 & 0 \\ 0 & \alpha_{22} & 0 \\ 0 & 0 & \alpha_{33} \end{pmatrix} \quad (8.74)
$$

where $\alpha_{11} = 8 \sum_{i=1}^{m}(\sum_{j=1}^{i-1} f_1(0) + g_1(0))$, $\alpha_{22} = 8 \sum_{i=1}^{m}(\sum_{j=1}^{i-1} f_2(0) + g_2(0))$, and $\alpha_{33} = 8 \sum_{i=1}^{m}(\sum_{j=1}^{i-1} f_3(0) + g_3(0))$. In addition, $f_{ll}(0)$ and $g_{ll}(0)$ are the maximum values of the pdfs of the components of

$$
\begin{pmatrix} 2(V_1 + V_2 + V_3 + V_4) \\ 2(V_1 + V_2 - V_3 - V_4) \\ 2(V_1 - V_2 + V_3 - V_4) \end{pmatrix} \quad (8.75)
$$

and

$$
\begin{pmatrix} (V_1 + V_2 + V_3 + V_4) \\ (V_1 + V_2 - V_3 - V_4) \\ (V_1 - V_2 + V_3 - V_4) \end{pmatrix} \quad (8.76)
$$

Therefore, the diagonal elements can be written as

$$
\alpha_{ll} = 8[m(m-1)f_{ll}(0) + mg_{ll}(0)] \qquad l = 1, 2, 3 \quad (8.77)
$$

Because the symmetric pdf of the vector sequence V_k may not be known, it is required to find an estimate of both $f_{ll}(0)$ and $g_{ll}(0)$ to completely determine the elements of the gain matrix. Kadar and Kurz[39] show that the diagonal elements α_{ll} can be approximated as

$$
\alpha_{ll} \simeq 8m^2 g_{ll}(0) \qquad l = 1, 2, 3 \quad (8.78)
$$

where use has been made of the generalized Gaussian noise representation

$$g^c(x) = \prod_{l=1}^{3} g_l^c(x) \tag{8.79}$$

with $g_l^c(x) = [c/2A(c)\Gamma(1/c)]\exp\{-[|x_l|/A(c)]^c\}$, $l = 1, 2, 3$ and $A(c) = [\sigma_l^2\Gamma(1/c)\Gamma(3/c)]^{1/2}$. The covariance of $g^c(x)$ is diagonal with elements σ_l^2, $l = 1, 2, 3$. Finally, the elements of the adaptive gain matrix are obtained by deriving a robust estimator of $[8m^2 g_{ll}(0)]^{-1}$. It can be shown[39] that

$$[A_k]_{ll} = \frac{m+1}{8(k-1)} \sum_{j=1}^{k-1} [Z_{j,[m/2]+1}^{ll} - Z_{j,[m/2]}^{ll}] \qquad l = 1, 2, 3 \tag{8.80}$$

where $Z_{j,[m/2]+1}^{ll} \equiv \{[m/2]+1\} - th$ order statistic from the component of random samples $Y_{i+m(k-i)}^l$, $i = 1, 2, \ldots, m$; $l = 1, 2, 3$ with pdf $g(x) = \prod_{l=1}^{3} g_l(x)$. Here, $[m/2]$ is defined to be the greatest integer less than or equal to $m/2$.

Finally, a more flexible and extendable to dependent sampling approach to robustizing the stochastic approximation algorithm is based on the concept of M-interval partition approximation (MIPA) preprocessor suggested by Tsai and Kurz.[75, 76] An extension of this theory to dependent sampling based on the strongly mixing model was developed by one of the senior author's students.[82] Though the underlying proofs and theory may be complicated, the actual algorithm is simple, efficient, and flexible.

8.6 Power calculations

In this section we calculate the probability of detection corresponding to the detectors based on F-tests and contrast tests, respectively. For each case, the independent and dependent noise cases are considered. Without loss of generality, we assume that the detection is based on one test statistic only.

8.6.1 F-statistic based detectors

A large class of detectors presented in this book are based on F-statistics. The methodology is as follows. First, an effect from a p-way design is coupled with a physical feature such as line, edge, etc. Then, the calculated F-statistic is compared to a tabulated threshold. Fig. 8.3 shows the block diagram of the procedure.

The power of the F-test is written as

$$P_D = Pr\left[F_{\nu_1,\nu_2,\delta} > F'_{\alpha,\nu_1,\nu_2}\right] \tag{8.81}$$

where δ is the noncentral parameter and α is the significance level. The probability of detection is dependent on the degrees of freedom ν_1, ν_2, α, and δ. Assume we

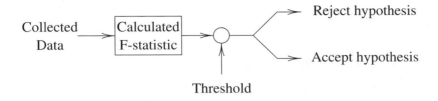

Figure 8.3. F-statistic based detector block diagram.

have selected a design with a given number of ways of yielding ν_1 and ν_2. Then, P_D can be found by using the tabulated charts of Pearson and Hartley. In this case, instead of δ, we use ϕ defined as follows

$$\phi = \delta(\nu_1 + 1)^{-1/2} \tag{8.82}$$

Example   Consider a one-way design with a 5×5 scanning mask over a 3-pixel wide edge. Using Scheffe's rule for finding δ,[2] we have

$$\sigma^2\delta^2 = \nu_1 \sum_{j=1}^{\nu_2} \beta_j^2 \tag{8.83}$$

where β denotes column effects.

Assume that the pixels are uniform in value with edge and background pixels assigned the values g_e and g_b, respectively (see Fig. 8.4). The parameter estimates are

$$\begin{aligned}
\widehat{\mu} &= \tfrac{3}{5} g_e \\
\beta_1 = \beta_2 &= \tfrac{-3}{5} g_e \\
\beta_3 = \beta_4 = \beta_5 &= \tfrac{2}{5} g_e
\end{aligned} \tag{8.84}$$

which implies that $\phi = d/\sigma$. For a significance level $\alpha = .95$ and $P_D = .90$, the value of ϕ from the table is 2.00.[4]

g_b	g_b	g_e	g_e	g_e
g_b	g_b	g_e	g_e	g_e
g_b	g_b	g_e	g_e	g_e
g_b	g_b	g_e	g_e	g_e
g_b	g_b	g_e	g_e	g_e

Figure 8.4. 3-pixel wide edge.

In the dependent case, it is not possible to derive a closed form of the probability of detection. Instead, we rely on a Monte-Carlo simulation to obtain P_D. As shown

---

[4] $\nu_1 = 4$ and $\nu_2 = 20$.

by several studies,[10, 14, 66] P_D exhibits a V-curve behavior typical of the dependent noise algorithms in this book.

8.6.2 Contrast-based detectors

Using the contrast approach, we form a linear combination of the effects representing the physical feature and compare it to a tabulated threshold. Fig. 8.5 shows the block diagram of a typical detection procedure using a contrast function.

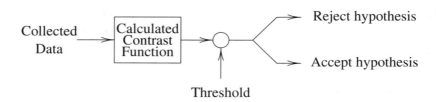

Figure 8.5. Contrast-based detector block diagram.

The hypothesis $H : \psi = 0$ is rejected at the α-significance level if

$$|\widehat{\psi}_i| > (q \, F_{\alpha, q, \nu_2})^{1/2} \sigma_{\widehat{\psi}_i} \qquad i = 1, 2, \ldots, q. \tag{8.85}$$

where we assume we have q contrast functions, $\widehat{\psi}_i$ denotes the estimate of the ith contrast and $\sigma_{\widehat{\psi}_i}$ is its corresponding variance estimate.

Eq. (8.85) can be rewritten in the following form

$$P_r\left[|\widehat{\psi}_i| > S\sigma_{\widehat{\psi}_i} / \text{ no contrast is present}\right] = \alpha \tag{8.86}$$

where $S^2 = q \, F_{\alpha, q, \nu_2}$.

Eq. (8.86) enables the introduction of the Neyman–Pearson detector with the following hypothesis-decision pair.

$$\begin{aligned} H : &\ \psi_i = 0 \text{ and } \psi_i \text{ is } N(0, \sigma_{\psi_i}^2) \\ K : &\ \psi_i \neq 0 \text{ and } \psi_i \text{ is } N(\psi_i, \sigma_{\psi_i}^2) \end{aligned} \tag{8.87}$$

The probability of detection can now be calculated (see Fig. 8.6). We have

$$P_D = \frac{1}{\sqrt{2\pi}\sigma_{\widehat{\psi}}} \int_{S\sigma_{\widehat{\psi}}}^{\infty} \exp \frac{-1}{2\sigma_{\widehat{\psi}}^2} (x - \psi)^2 dx \tag{8.88}$$

Since S depends only on the number of contrasts involved in the detection, the only parameter that affects P_D is the contrast variance $\sigma_{\widehat{\psi}_i}$.

In the dependent noise case, $\sigma_{\widehat{\psi}_i}$ is a function of the row and column correlation factors. A Monte-Carlo simulation is necessary for the calculation of P_D.

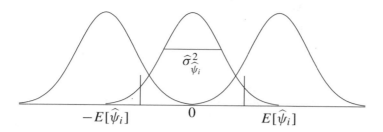

Figure 8.6. Power calculations.

8.7 Concluding remarks

The theory of SA applied to least squares estimation problems was presented in this chapter. After introducing contraction mapping in the deterministic case, we presented Dvoretzky's theorem, which provides a set of sufficient conditions to ensure convergence in the mean square sense and with probability one of a given iterative scheme.

Two well-known SA procedures, Robbins–Monro and Kiefer–Wolfowitz, were briefly discussed. In particular, we highlighted the Robbins–Monro procedure and its convergence because of its central role in RMSA techniques in least squares estimation.

Next, we presented the scalar and vector versions of Gladyshev's theorems. For both cases, the calculation of the optimum gain coefficients in the iterative scheme was highlighted. The bias inherent to small sample size was also demonstrated.

Robust recursive estimation techniques were introduced in the following section. Distribution-free statistics, both one-sample and two-sample, were discussed along with the LLIF.

Power calculations were developed for both the F-test and contrast-test based detectors. In the dependent noise case, it was shown via appropriate Monte-Carlo simulations that the probability of detection follows a V-type behavior with the minimum around .45 for the correlation factor.[5]

Appendix

Some distributions

The test statistics used throughout this book are based essentially on the F- and χ^2-distributions. We present in this section a brief review of both distributions.

---

[5]See references [10], [14], and [66].

Chi-square distribution

Let $z_1, z_2, \ldots, z_\nu$ be ν independent normal variates with $E[z_i] = \xi_i$. Then,

$$\chi_{\nu,\delta} = \sum_{i=1}^{\nu} z_i^2 \tag{A.1}$$

has a chi-square distribution where ν is the corresponding degree of freedom. The noncentrality parameter is $\delta = \left(\sum_{i=1}^{\nu} \xi_i^2 \right)^{1/2}$, and the mean of the distribution is $E[\chi^2] = \nu$. When $\delta = 0$, $\chi_{\nu,\delta=0}^2$ is called a central chi-square and denoted by χ_ν^2.

F-distribution

Let z_1 and z_2 be two independent random variables with distribution $\chi_{\nu_1,\delta}^2$ and $\chi_{\nu_2}^2$, respectively. Then, the quotient

$$F'_{\nu_1,\nu_2,\delta} = \frac{z_1/\nu_1}{z_2/nu_2} \tag{A.2}$$

has a noncentral F-distribution with degrees of freedom ν_1 and ν_2 with noncentrality parameter δ. When $\delta = 0$, the central variable $F'_{\nu_1,\nu_2,\delta=0}$ is denoted by F_{ν_1,ν_2}.

9

Some approaches to image restoration

9.1 Introductory remarks

In this chapter, two different classes of image restoration approaches are presented. Though the stress is placed on images corrupted by so-called "salt and pepper" noise, represented by the mixture distribution model,[73] the methodologies presented here are applicable if the background noise deviates significantly from Gaussian. The image restoration is accomplished in two stages. In the first stage, edge detectors introduced in Chapter 4 are used as preprocessors to establish the local orientation of potential edge points. In the second stage, some form of Robbins–Monro type recursive estimator (see Chapter 8) is applied to remove the undesirable corruption of the image. Alternately, the badly corrupted pixels are replaced by estimated values based on the missing value approach.[4] Based on extensive simulation studies in various noise environments, the edge detection preprocessors that were found to be of practical use are 5×5 Graeco-Latin squares (GLS) (see Section 4.5.2) and 6×7 Youden squares (YS) (see Section 3.7.2).

Many image restoration procedures, such as averaging and median filters, represent a smoothing process and will cause blurring of the restored image. The averaging filter represents a sample mean and is not robust in salt and pepper noise because of the high variance of the latter. Though the median filter suppresses the heavy-tailed section of the noise component, thereby causing an increase in the graininess of the image, its ability to deal with Gaussian noise is rather poor.[102] On the other hand, the procedures introduced in this chapter are flexible, robust, and computationally efficient, yielding restored images of excellent visual quality.

9.2 Edge detection preprocessors

As indicated in the previous section, the two-stage approach to image restoration uses an edge detector in the first stage to preserve the true edges. Though almost any edge processor can be used in this stage, experience with many corrupted

181

images indicates that the use of GLS and YS masks yields best overall results. The use of YS masks in the first stage leads to a better quality image, but the increase in the computation time sometimes does not justify the choice of YS mask as a preprocessor. In most applications the GLS masks produce acceptable results.

An outline of the procedure for a 7×6 YS mask with Markov dependent noise is presented next. For the details involving a 5×5 GLS mask one should refer to Chapters 2 and 4.

A digitized image may be modeled as a two-dimensional intensity pixel function $f(i, j)$. Let $y(i, j)$ be a pixel value in a localized mask.

Mathematically, the model may be described (see Section 2.5.3) by

$$y_{ijlh} = \mu + \alpha_i + \beta_j + \sum_{l}^{k} \tau_l n(i, j, l) + \sum_{h}^{k} \lambda_h m(i, j, h) + e_{ijlh} \qquad (9.1)$$

where $n(i, j, l) = 1$ if treatment τ_l is present in cell (i, j) and $m(i, j, h) = 1$ if treatment λ_h is presented in cell (i, j); otherwise they are zero, respectively. α_i, $i = 1, \ldots, t$, are the row effects; β_j, $j = 1, \ldots, b$, are the column effects; τ_l, $l = 1, \ldots, k$, are treatments represented by Latin letters; and λ_h, $h = 1, \ldots, k$, are treatment effects represented by Greek letters.

This model enables us to determine the presence of row, column, Latin, and Greek effects, namely four directions $(0°, 45°, 90°, 135°)$, simultaneously. The model chosen is a special case of a Youden design with

$$b \times k = t \times \rho \quad \text{and} \quad \omega = \frac{\rho(I - 1)}{b - 1} \qquad (9.2)$$

where ρ is the number of treatment replicates.

The number of blocks is b and is equal to the number of treatment k. In this case $\rho = t$, and ω is the number of times a pair of treatments appears together in the design. For the mask under consideration, ω_t and ω_l are the same. To simplify computation, the following mask is chosen (the T-set treatments is the set of treatment effects represented by the Latin letters, and the Λ-set is the set of treatments represented by Greek letters).

$$
\begin{array}{ccccccc}
C : \alpha & D : \beta & E : \nu & F : \delta & G : \varepsilon & A : \varphi & B : \gamma \\
D : \gamma & E : \alpha & F : \beta & G : \nu & A : \delta & B : \varepsilon & C : \varphi \\
E : \varphi & F : \gamma & G : \alpha & A : \beta & B : \nu & C : \delta & D : \varepsilon \\
F : \varepsilon & G : \varphi & A : \gamma & B : \alpha & C : \beta & D : \nu & E : \delta \\
G : \delta & A : \varepsilon & B : \varphi & C : \gamma & D : \alpha & E : \beta & F : \nu \\
A : \nu & B : \delta & C : \varepsilon & D : \varphi & E : \gamma & F : \alpha & G : \beta
\end{array}
$$

The estimates of α_i, β_j, τ_l, and λ_h under the alternative are found via the LMSE procedure, which minimizes

$$S(y, B) = \sum_{i=1}^{t} \sum_{j=1}^{b} \left(y_{ijlh} - \mu - \alpha_i - \beta_j - \sum_{h=1}^{k} \lambda_h m(i, j, h) \right)^2, \ N = t \cdot b. \quad (9.3)$$

Without any loss of generality, the linear mathematical model can be written for the alternative space as $\mathbf{Y} = \mathbf{X}^T \mathbf{B} + \mathbf{e}$. Where $\mathbf{X}^{p \times N}$ is the design matrix with $r = \text{rank}(\mathbf{X})$, $\mathbf{B}^{p \times 1}$ is the matrix of factors, $[\underline{\alpha} \ \underline{\beta} \ \underline{\tau} \ \underline{\lambda}]^t$ to be estimated, and $\mathbf{e}^{N \times 1}$ is the noise component matrix. We test the hypothesis $H_i : f^i = 0$ under w_i while the space under the alternative Ω tests for some f^i are different from zero with

$$f^1 = \alpha_i \qquad f^2 = \beta_j \qquad f^3 = \tau_l \qquad f^4 = \lambda_h$$

The sum of squares of errors under the alternative and hypothesis H_i are, respectively

$$S_\Omega = (\mathbf{Y} - \mathbf{X}^T \widehat{\mathbf{B}})^T (\mathbf{Y} - \mathbf{X}^T \widehat{\mathbf{B}}) \qquad\qquad (9.4)$$

and

$$S_{w_i} = (\mathbf{Y} - \mathbf{X}^T \widehat{\mathbf{B}}_{w_i})^T (\mathbf{Y} - \mathbf{X}^T \widehat{\mathbf{B}}_{w_i}) \qquad\qquad (9.5)$$

The test statistic

$$F_i = \frac{N - r}{q_i} \frac{S_{w_i} - S_\Omega}{S_\Omega}$$

is to be compared to $F_{\alpha, q_i, N-r}$. If $F_i \geq F_{\alpha, q_i, N-r}$, we reject the hypothesis H_i with a confidence level α.

$\widehat{\mathbf{B}}$ and $\widehat{\mathbf{B}}_{w_i}$ can be found using the LS estimation procedure under the alternative and hypothesis, respectively, and q_i is the degree of freedom of the factor effects f^is.

Following the same procedure as that presented in Chapter 2 and for the Markov dependent case, it is assumed that $E[\mathbf{e}\mathbf{e}^T] = \sigma^2 \mathbf{K}_f$ where $\mathbf{K}_f$ is the covariance matrix. It is also assumed that the noise is wide sense stationary and can be modeled by a tensored form matrix or $\mathbf{K}_f = \mathbf{K}_r \otimes \mathbf{K}_c$. $\mathbf{K}_r$ and $\mathbf{K}_c$ are, respectively, the row and column correlation matrices. The tensored form means that the correlation among pixels is separable as the product of row and column correlation functions. This type of dependency of the noise component is also known as dependent Markov noise resulting in

$$\mathbf{K}_c{}^{b \times b} = \sigma_c^2 \begin{pmatrix} 1 & b & b^2 & \dots & b^{m-1} \\ b & 1 & b & \dots & b^{m-2} \\ \vdots & \vdots & \vdots & \ddots & \vdots \\ b^{m-1} & b^{m-2} & b^{m-3} & \dots & 1 \end{pmatrix} \qquad (9.6)$$

and

$$\mathbf{K}_r{}^{t \times t} = \sigma_r^2 \begin{pmatrix} 1 & a & a^2 & \dots & a^{n-1} \\ a & 1 & a & \dots & a^{n-2} \\ \vdots & \vdots & \vdots & \ddots & \vdots \\ a^{n-1} & a^{n-2} & a^{n-3} & \dots & 1 \end{pmatrix} \tag{9.7}$$

Therefore, the Kroenecker product yields

$$\mathbf{K}_f{}^{N \times N} = \sigma^2 \begin{pmatrix} K_{r_{11}} (\mathbf{K}_c) & K_{r_{12}} (\mathbf{K}_c) & \dots & K_{r_{1t}} (\mathbf{K}_c) \\ \vdots & \vdots & \ddots & \vdots \\ K_{r_{t1}} (\mathbf{K}_c) & K_{r_{t1}} (\mathbf{K}_c) & \dots & K_{r_{tt}} (\mathbf{K}_c) \end{pmatrix} \tag{9.8}$$

If $P(i_p, j_p)$ and $Q(i_q, j_q)$ are two points in the image, their correlation is given by $a^{|i_p - i_q|} b^{|j_p - j_q|}$, a and b being the correlation coefficient for rows and columns, respectively. To obtain the uncorrelated data set, one must transform the data through an orthogonal linear transformation. Because $\mathbf{K}_f$ is a symmetric positive definite matrix, there exits a nonsingular matrix $\mathbf{P}^{N \times N}$ such that $\mathbf{P}^T \mathbf{K}_f \mathbf{P} = \mathbf{I}$. As indicated in Chapter 2, $\mathbf{P}$ is the matrix of eigenvalues of $\mathbf{K}_f$ that diagonalizes $\mathbf{K}_f$. One transformation that might be used is based on the Cholesky factorization,[10] which states that if $\mathbf{P}^T$ is a lower triangular matrix, then $\mathbf{P}^T \mathbf{P} = \mathbf{K}_f$. This choice leads to $\mathbf{P} \mathbf{P}^T = \mathbf{K}_f{}^{-1}$. If we apply $\mathbf{P}$ to the correlated data $\mathbf{y}$, we obtain a new set of uncorrelated data points $\tilde{\mathbf{y}} = \mathbf{P}^T \mathbf{y}$. By taking the expectation, we obtain

$$E[\tilde{\mathbf{y}} \tilde{\mathbf{y}}^T] = \mathbf{P}^T E[\tilde{\mathbf{y}} \tilde{\mathbf{y}}^T] \mathbf{P} = \mathbf{P}^T \sigma^2 \mathbf{K}_f \mathbf{P} = \sigma^2 \mathbf{I} \tag{9.9}$$

So $\tilde{\mathbf{y}}$ is $N(\mathbf{P}^T \mathbf{X}^T \mathbf{B}, \sigma^2 \mathbf{I})$. The sum of square error is then

$$S(\tilde{\mathbf{y}}, B) = (\tilde{\mathbf{y}} - \mathbf{P}^T \mathbf{X}^T \mathbf{B})^T (\tilde{\mathbf{y}} - \mathbf{P}^T \mathbf{X}^T \mathbf{B}) = (\mathbf{y} - \mathbf{X}^T \widehat{\mathbf{B}})^T \mathbf{K}_f{}^{-1} (\mathbf{y} - \mathbf{X}^T \widehat{\mathbf{B}}) \tag{9.10}$$

The unbiased estimate of the variance σ^2 is in this case

$$s^2 = \frac{(\mathbf{y} - \mathbf{X}^T \widehat{\mathbf{B}})^T \mathbf{K}_f{}^{-1} (\mathbf{y} - \mathbf{X}^T \widehat{\mathbf{B}})}{N - r} \tag{9.11}$$

The hypothesis testing and the statistical tests can be formulated in the same way as the uncorrelated case.

Let $Y_{ijlh}^e = y_{ijlh} - \mu - \beta_j - \sum_l n(i, j, l) \tau_l - \sum_h m(i, j, h) \lambda_h = e_{ijlh}$, $i = 1, \dots, l, j = 1, \dots, b, l = 1, \dots, k, h = 1, \dots, k$.

It can be shown that $S(Y^e, B) = E(Y^e Y^{e^t})$ has the following form

$$S(Y^e, B) = \sum_i \sum_j \sum_q \sum_p R_{iq} C_{jp} y_{qp}^e y_{ijlh}^e,$$

$$q = 1, \dots, t, \quad p = 1, \dots, b.$$

The side conditions are

$$\sum_i R_{i.}\alpha_i = 0, \quad \sum_j C_{j.}\beta_j = 0 \tag{9.12}$$

$$\sum_i \sum_j \sum_l R_{i.}C_{j.}n(i,j,l)\tau_l = 0, \quad \sum_i \sum_j \sum_h R_{i.}C_{j.}m(i,j,l)\lambda_h = 0 \tag{9.13}$$

where $R_{..} = \sum_i \sum_q R_{iq}$, $R_{i.} = \sum_q R_{iq}$, $C_{..} = \sum_j \sum_p C_{jp}$, and $C_{j.} = \sum_p C_{jp}$.

Through minimization, the estimates of the factor effects under the alternative space are solutions of the set of equations

$$\sum_i \sum_j R_{i.}C_{j.}y_{ij..} = R_{..}C_{..}\mu \tag{9.14}$$

$$\sum_q \sum_p R_{iq}C_{p.}y_{qplh} = C_{..}R_{i.}\mu + C_{..}\sum_q R_{iq}\alpha_q$$
$$+ \sum_q \sum_p \sum_l R_{iq}C_{p.}\tau_l n(q,p,l)$$
$$+ \sum_q \sum_p \sum_h R_{iq}C_{p.}\lambda_h m(q,p,h) \tag{9.15}$$

$$\sum_q \sum_p R_{q.}C_{jp}y_{qplh} = R_{..}C_{j.}\mu + R_{..}\sum_p C_{jq}\beta_p$$
$$+ \sum_q \sum_p \sum_l R_{q.}C_{jp}\tau_l n(q,p,l)$$
$$+ \sum_q \sum_p \sum_h R_{q.}C_{jp}\lambda_h m(q,p,h) \tag{9.16}$$

$$\sum_i \sum_j \sum_q \sum_p n(i,j,l)R_{iq}C_{jp}y_{qpl'h'}$$
$$= \sum_i \sum_j n(i,j,l)R_{i.}C_{j.}\mu + \sum_i \sum_j \sum_q n(i,j,l)R_{iq}C_{j.}\alpha_q$$
$$+ \sum_i \sum_j \sum_p n(i,j,l)R_{i.}C_{jp}\beta_p$$
$$+ \sum_i \sum_j \sum_q \sum_p \sum_{l'} n(i,j,l)n(q,p,l')R_{iq}C_{jp}\tau_{l'}$$
$$+ \sum_i \sum_j \sum_q \sum_p \sum_{h'} n(i,j,l)m(q,p,h')R_{iq}C_{jp}\lambda_{h'} \tag{9.17}$$

$$\sum_i \sum_j \sum_q \sum_p m(i, j, h) R_{iq} C_{jp} y_{qpl'h'}$$

$$= \sum_i \sum_j m(i, j, h) R_{i.} C_{j.} \mu + \sum_i \sum_j \sum_q m(i, j, h) R_{iq} C_{j.} \alpha_q$$

$$+ \sum_i \sum_j \sum_p m(i, j, h) R_{i.} C_{jp} \beta_p$$

$$+ \sum_i \sum_j \sum_q \sum_p \sum_{l'} m(i, j, h) n(q, p, l') R_{iq} C_{jp} \tau_{l'}$$

$$+ \sum_i \sum_j \sum_q \sum_p \sum_{h'} m(i, j, h) m(q, p, h') R_{iq} C_{jp} \lambda_{h'} \qquad (9.18)$$

Finally, the normal Eqs. (9.14) through (9.18) can be written in matrix form and grouped as

$$\begin{pmatrix} aa & ab & at & al & am \\ ba & bb & bt & bl & bm \\ ta & tb & tt & tl & tm \\ la & lb & lt & ll & lm \\ ma & mb & mt & ml & mm \end{pmatrix} \begin{pmatrix} \underline{\alpha} \\ \underline{\beta} \\ \underline{\tau} \\ \underline{\lambda} \\ \mu \end{pmatrix} = \begin{pmatrix} ay \\ by \\ ty \\ ly \\ my \end{pmatrix} \mathbf{Y} \qquad (9.19)$$

or

$$\mathbf{WB} = \mathbf{ZY} \qquad (9.20)$$

The matrix $\mathbf{W}$ is usually nonsingular; therefore, the estimates of $\mathbf{B}$ are given by $\widehat{\mathbf{B}} = (\mathbf{W}^{-1}\mathbf{Z})\mathbf{Y}$. We note that the matrices $\mathbf{W}$ and $\mathbf{Z}$ do not depend on the observations but on the correlation matrix only. Moreover, if we assume that the entire image has the same correlation matrix, then $\mathbf{W}^{-1}\mathbf{Z}$ is computed only once. This assumption usually results in a drastic reduction in computation time.

9.3 Image restoration using linear regressions

As a next stage in the edge-preserving restoration, one can fit a linear regression model to the noisy image. Fitting a simple or a multiple linear regression model performs better than just averaging or median filtering. Parameters of the multiple linear regression model are then estimated by a recursive least square estimator of the type described in Chapter 8.

For a two-dimensional image a multiple linear regression model is of the form

$$y_{ij} = b_0 + b_1 x_{1i} + b_2 x_{2j} + e_{ij}, \quad j = 1, \dots, n, \quad i = 1, \dots, k \qquad (9.21)$$

where b_0, b_1, and b_2 are the parameters of the multiple linear regression to be estimated from the noisy observations.

In matrix form, Eq. (9.21) may be written as

$$\mathbf{Y} = \mathbf{Xb} + \mathbf{e} \tag{9.22}$$

where $\mathbf{Y} = [y_1, y_2, \ldots, y_n]^T$, $\mathbf{b} = [b_0, b_1, b_2]$, $\mathbf{e} = [e_1, e_2, \ldots, e_n]$.

The matrix $\mathbf{X}$ is referred to as the "design matrix" (see Chapter 2). For a 5×5 GLS mask, matrix $\mathbf{X}$ is given by

$$\mathbf{X} = \begin{pmatrix} 1 & -2 & -2 \\ 1 & -2 & -1 \\ 1 & -2 & 0 \\ 1 & -2 & +1 \\ 1 & -2 & +2 \\ 1 & -1 & -2 \\ 1 & -1 & -1 \\ 1 & -1 & 0 \\ 1 & -1 & +1 \\ 1 & -1 & +2 \\ 1 & 0 & -2 \\ 1 & 0 & -1 \\ 1 & 0 & 0 \\ 1 & 0 & +1 \\ 1 & 0 & +2 \\ 1 & +1 & -2 \\ 1 & +1 & -1 \\ 1 & +1 & 0 \\ 1 & +1 & +1 \\ 1 & +1 & +2 \\ 1 & +2 & -2 \\ 1 & +2 & -1 \\ 1 & +2 & 0 \\ 1 & +2 & +1 \\ 1 & +2 & +2 \end{pmatrix} \tag{9.23}$$

which yields

$$\mathbf{X}^T\mathbf{X} = \begin{pmatrix} 25 & 0 & 0 \\ 0 & 50 & 0 \\ 0 & 0 & 50 \end{pmatrix} \tag{9.24}$$

and by inversion

$$(\mathbf{X}^T\mathbf{X})^{-1} = \begin{pmatrix} 1/25 & 0 & 0 \\ 0 & 1/50 & 0 \\ 0 & 0 & 1/50 \end{pmatrix} \tag{9.25}$$

Other choices for x_{1i}, x_{2j} are possible to yield $(\mathbf{X}^T\mathbf{X}) = [\mathbf{I}]_{3\times 3}$.

If only a row edge is detected between the first and second rows or the fourth and fifth rows by the GLS mask and the F-test is performed on pair of means, then $x_{1i} = -1, -1, +1, +3$ and $x_{2j} = -2, -1, 0, +1, +2$ and a similar $\mathbf{X}$ matrix is formed having twenty rows.

If only a row edge is detected between the second and third rows or the third and fourth rows, then the choice of x_1, x_2 is $x_{1i} = -1, 0, +1$ and $x_{2j} = -1, -1, 0, +1, +2$. Thus, the design matrix $\mathbf{X}$ will have 15 rows.

Similarly, $\mathbf{X}$ matrices are formed for only a column edge or row and column edges. For discriminating diagonal edges, a simple linear regression model will be of the form

$$y_i = b_0 + b_1 x_1 + e \tag{9.26}$$

where $x_1 = -2$, $x_2 = -1$, $x_3 = 0$, $x_4 = +1$, and $x_5 = +2$. This leads to

$$\mathbf{X} = \begin{pmatrix} 1 & -2 \\ 1 & -1 \\ 1 & 0 \\ 1 & 1 \\ 1 & 2 \end{pmatrix} \tag{9.27}$$

Hence,

$$\mathbf{X}^T\mathbf{X} = \begin{pmatrix} 5 & 0 \\ 0 & 10 \end{pmatrix} \tag{9.28}$$

and its inverse being

$$(\mathbf{X}^T\mathbf{X})^{-1} = \begin{pmatrix} 1/5 & 0 \\ 0 & 1/10 \end{pmatrix} \tag{9.29}$$

Other choices for x_i are possible by normalizing $\mathbf{X}$. This leads to $(\mathbf{X}^T\mathbf{X}) = [\mathbf{I}]_{3\times 3}$.

Now that we have defined the design matrix, it remains to find the parameter vector. Toward that end, a recursive least square estimator of the type discussed in Chapter 8 is represented by

$$\mathbf{b}_{k+1} = \mathbf{b}_k + \frac{A_k}{k}(\mathbf{X}^T\mathbf{X})^{-1}\mathbf{X}^T\mathbf{Z}_k \tag{9.30}$$

where

$$\mathbf{Z}_k = \mathbf{Y} - \mathbf{X}\mathbf{b}_k \tag{9.31}$$

and its convergence to the appropriate $\mathbf{b}$ can be proven as outlined in the previous chapter.

If the noise is severely contaminated by a high-variance component, the estimator of Eq. (9.30) can be robustized as shown in Section 8.4.1 by using a rank statistic preprocessor of the Wilcoxon type.

Variations on least square procedures outlined above have been used success-fully. If the high-variance contamination is small, a simple procedure suggested in references [99] and [100] is useful. If the noise is poorly defined and/or slowly varying, a class of modified robust image processors[72, 73, 101] performs better than the processors suggested in this section.

9.4 Suppression of salt and pepper noise

If the non-Gaussian noise is confined to the salt and pepper type, a simple second stage procedure for image reconstruction based on the so-called missing value approach[4] may be devised. The salt and pepper noise is mathematically described by the mixture distribution model with pdf.

$$f_e(x) = (1 - \varepsilon)f_n(x) + \varepsilon f_i(x) \quad 0 \le \varepsilon \le 1 \tag{9.32}$$

where $f_n(x)$ is the normal pdf, usually Gaussian, and $f_i(x)$ is a pdf of high variance representing impulsive noise. It should be noted that even if the impulsive noise is high-variance Gaussian, the mixture noise is no longer Gaussian. The model given in Eq. (9.32) assumes uncorrelated noise. If $f_n(x)$ is Markov Gaussian, then it is $N(0, \sigma_f^2 \mathbf{K}_f)$ where $\mathbf{K}_f$ was defined earlier in the chapter. The remaining part of the noise model remains the same. It is also assumed that $f_n(x)$ and $f_i(x)$ are statistically independent.

If the preprocessing mask is of the YS type, we insert the missing value (severely corrupted pixels) $Z(i_z, j_z)$ into the sum of square errors (SSE) S_Ω, and write S_Ω as a sum of two components. One component involves the z-terms only, and the other accounts for the remaining terms, namely

$$S_\Omega = S(Y_{obs}, B) + S(Z, B) \tag{9.33}$$

where Y_{obs} is the set of the observations minus the missing values pixels. We seek an estimate for Z that will yield a minimum error sum of squares. Solving for $\partial S_\Omega / \partial Z = 0$, yields the estimated value that is then incorporated into the data to obtain a complete data, and ANOVA is applied with a change in the degrees of freedom because we have only $n - 1$ observations. If n_z points are considered missing, a recursive estimation of these points is used, and the degrees of freedom for the residual sum of squares and the total sum of squares are, respectively, $n_e - n_z$ and $N - n_z$, while the degree of freedom for the SSE is n_e. The reader may refer to Cox and Cochran[4] for the treatment of missing data in various experimental designs.

In this section new formulas for a four-way Youden design are derived for the uncorrelated data case. If several observations are missing, we use a recursive procedure to estimate them rather than differentiate the error sum of square with respect to each one of them, and solving the normal equations would be just as laborious. The recursive procedure for two missing variables, let us say z_1 and z_2, is to estimate z_1 via the LSE and insert the estimate of z_1 in the data to estimate z_2. Insert the estimate of z_2 in the data to reestimate z_1, which in turn is inserted in the data to reestimate z_2. This recursive method is repeated until $|z_i^{n+1} - z_i^n| < \xi$ where ξ is predetermined by the experimenter. Let $\mathbf{y}^{n \times 1} = \mathbf{X}^t \mathbf{B} + \mathbf{e}$ with missing observation z_k.

We can write the SSE for the complete data as

$$S_{com}(y, B) = S_{incomp}(y, b) + \|z_k - x_k^T \mathbf{B}\|^2 \tag{9.34}$$

where S_{incomp} is the SSE for the incomplete value data. The estimate for the missing observation is the solution of

$$\frac{\partial \|z_k - x^t \widehat{B}\|^2}{\partial z_k} = 0$$

Correlated Data—Least Square Estimate of the Missing Value
The error sum of square can be written as

$$S(Y^e, B) = \sum_i \sum_j \sum_p \sum_q R_{iq} C_{jp} y_{ij}^e y_{qp}^e \tag{9.35}$$

where $y_{ij}^e = y_{ij} - \mu - \alpha_i - \beta_j - \sum_l n(i, j, l)\tau_l - \sum_h m(i, j, h)\lambda_h$, R_{iq} is the correlation coefficient between rows i and q, and C_{ip} is the correlation coefficient between columns j and p. Let $z = z(i_z, j_z)$ be the missing observation and k_z and l_z be, respectively, the Latin and Greek treatment assigned to cell (i_z, j_z). We insert z into the error sum of squares $S(Y^e, B)$, obtaining

$$
\begin{aligned}
S(Y^e, B, z^e) = {} & \sum_i \sum_j \sum_q \sum_p R_{iq} C_{jp} y_{ij}^e y_{qp}^e \Bigg|_{\substack{(i,j) \neq (i_z, j_z) \\ (q,p) \neq (i_z, j_z)}} \\
& + 2 \sum_q \sum_p R_{i_z, q} C_{j_z, p} z_{i_z, j_z}^e y_{qp}^e \Bigg|_{(q,p) \neq (i_z, j_z)} \\
& + R_{i_z, i_z} C_{j_z, j_z} z_{i_z, j_z}^{e^2}
\end{aligned} \tag{9.36}
$$

We are seeking the value of z_{i_z, j_z}^e and, therefore, z_{i_z, j_z}, that will result in the minimum of the error sum of squares, or we need to solve

$$\frac{\partial S(Y^e, B, z^e)}{\partial z^e} = 0$$

In the following equations $(q, p) \neq (i_z, j_z)$. By differentiating Eq. (9.36), we have

$$2R_{i_z, i_z} C_{j_z, j_z} z_{i_z, j_z}^e + 2 \sum_q \sum_p R_{i_z, q} C_{j_z, p} y_{qp}^e = 0 \tag{9.37}$$

which implies by replacing y^e

$$\sum_q \sum_p R_{i_z,q} C_{j_z,p} \left(y_{qp} - \mu - \alpha_q - \beta_p - \sum_l n(q,p,l)\tau_l \right.$$
$$\left. - \sum_h m(q,p,h)\lambda_h \right)$$
$$R_{i_z,i_z} C_{j_z,j_z} \left(z_{i_z,j_z} - \mu - \alpha_{i_z} - \beta_{j_z} - \sum_l n(i_z,j_z,l)\tau_l \right.$$
$$\left. - \sum_h m(i_z,j_z,h)\lambda_h \right) = 0 \tag{9.38}$$

Hence, the missing value is given by

$$z_{i_z,j_z} = \frac{-1}{R_{i_z,i_z} C_{j_z,j_z}} \left[\sum_q \sum_p R_{i_z,q} C_{j_z,p} y_{qp} - R_{i_{z'}} . C_{j_{z'}} . \mu - C_{j_{z'}} . \sum_q R_{i_{z'},q} \alpha_q \right.$$
$$- R_{i_{z'}} . \sum_p C_{j_z,p} \beta_p - \sum_q \sum_p R_{i_z,q} C_{j_z,p} n(q,p,l)\tau_l$$
$$\left. - \sum_q \sum_p R_{i_z,q} C_{j_z,p} m(q.p.h)\lambda_h \right] \tag{9.39}$$

An important task is to determine whether a pixel has been severely contaminated by impulsive noise and, therefore, must be considered as a pixel with a missing value. We arrange the N pixel values available from the mask in ascending order. Hence,

$$Y^T = (y_{11}, y_{12}, \ldots, y_{1b}, y_{21}, \ldots, y_{Ib})$$
$$\mapsto X^T = (x_1 = \min(y_{ij}), \ldots, x_N = \max(y_{ij})) \tag{9.40}$$

We divide the space generated by $\mathbf{X}^T$ into four intervals $Q_i; i = 1, \ldots, 4$.

Let $Q_1 = [a_1, a_2)$, $Q_2 = [a_2, a_3)$, $Q_3 = [a_3, a_4)$, and $Q_4 = [a_4, a_5]$ where $a_1 = x_1 = \min(y_{ij})$, $a_2 = (a_1 + a_3)/2$, $a_3 = \bar{x} = \sum_i^{ik} \sum_j^{ib} y_{ij}/N$, $a_4 = (a_3 + a_5)/2$, $a_5 = x_N = \max(y_{ij})$, and n_i the number of samples falling in Q_i.

One should note that in the absence of impulsive noise in the mask samples, if a segment of the line or edge is present, an appreciable fraction of the mask pixel values of the feature will fall into Q_4. If there is no feature (only noisy background), we can assume that n_is would not be significantly different from $N/4$. In the case where impulsive noise is present, the maximum of $|x_1|$ and/or $|x_N|$ is very large as compared to the rest of the pixel values. Impulsive noise-contaminated pixels would fall into Q_1 or Q_4. As a threshold, we choose:

If $n_1/N < \varepsilon_n = 0.05$, we have n_1 impulsive noise-contaminated pixels;
if $n_4/N < \varepsilon_n = 0.05$, we have n_4 impulsive noise-contaminated pixels.

Otherwise, we consider there is no pixel corrupted by impulsive noise.

If a missing pixel has been detected and located using this procedure, we find the estimate that would yield a minimum sum of squares of errors as shown previously. We insert this estimate in the data to obtain a complete data, and we carry out the ANOVA procedure with the appropriate degrees of freedom. If the F-test fails to reject all hypotheses, then, as a way of restoring the missing value pixel of the background, we replace it by the average mean of six out of eight surrounding neighbors, from which we reject the largest and smallest pixel values. If more than one pixel is considered as a member of the missing value pixel set within the mask, a recursive method to estimate the pixels is used. This approach requires small data storage and reasonable processing time.

9.5 Some results

To evaluate the relative merits of procedures suggested in this chapter, numerous simulation studies for both classes of procedures were undertaken. Both subjective and objective measures of performance were used. Subjective evaluations were based on the visual comparison of the simulation results, and objective evaluations were based on the improvement in the effective SNR.

For imaging data, the SNR is defined for the input as

$$SNR = 10 \log \frac{\sum_{i=1}^{ik} \sum_{j=1}^{ib} s_{ij}^2}{\sum_{i=1}^{ik} \sum_{j=1}^{ib} (y_{ij} - s_{ij})^2}$$

(a)

Figure 9.1. "Lena" image. (a) Noise contamination is $F(.) = 0.90N(0, 10^2) + 0.10N(0, 20^2)$. (b) Noisy image–frame 1, (c) noisy image–frame 2, (d) noisy image–frame 3, (e) noisy image–frame 4, (f) noisy image–frame 5, (g) restored image using a robustized Wilcoxon preprocessor estimator (iteration # 1000).   *(Continues)*

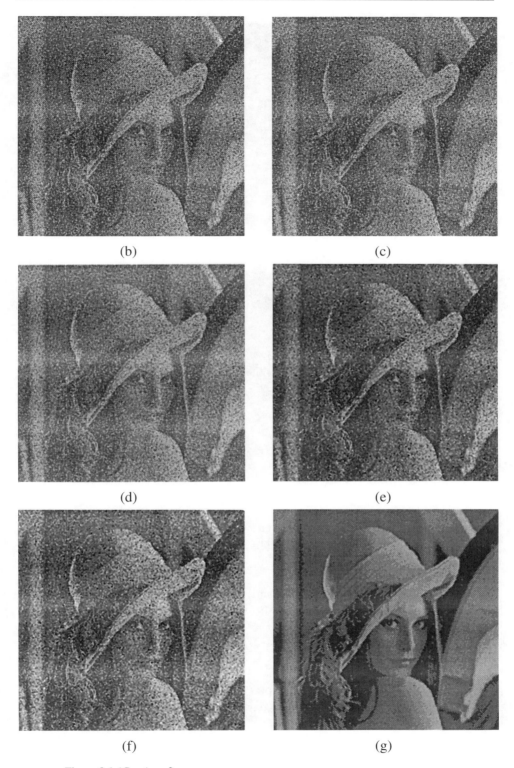

(b)

(c)

(d)

(e)

(f)

(g)

Figure 9.1 (*Continued*).

Input : SNR = 10.63 dB   Output: SNR = 16.24 dB

(a)

Input : SNR = 7.28 dB   Output : SNR = 15.56 dB

(b)

Figure 9.2. Image restoration based on modified Youden design. (a) Mixture noise:
$H(.) = 90\%N(0, 10^2) + 10\%N(0, 100^2)$, (b) mixture noise:
$H(.) = 90\%N(0, 5^2) + 10\%N(0, 50^2)$.

where s_{ij} is the original pixel value and y_{ij} is the corrupted pixel value at point
(i, j). To compare the input/output SNR, we define the output SNR as

$$SNR = 10 \log \frac{\sum_{i=1}^{ik} \sum_{j=1}^{ib} s_{ij}^2}{\sum_{i=1}^{ik} \sum_{j=1}^{ib} (\widehat{s}_{ij} - s_{ij})^2}$$

where $\widehat{s}_{ij}$ is the restored or estimated pixel.

For simulation purposes the impulsive noise is represented by a high-variance
Gaussian distribution.

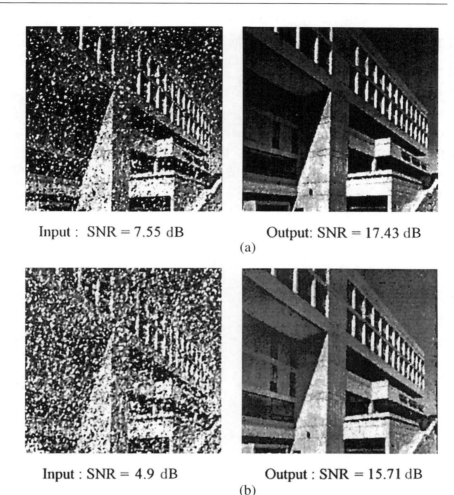

Input : SNR = 7.55 dB   Output: SNR = 17.43 dB

(a)

Input : SNR = 4.9 dB   Output : SNR = 15.71 dB

(b)

Figure 9.3. Image restoration based on modified Youden design in correlated noise.
(a) Mixture noise: $H(.) = 95\%N(0, 10^2) + 5\%N(0, 100^2)$, correlation coefficient $= 0.2$,
(b) mixture noise. $H(.) = 95\%N(0, 20^2) + 5\%N(0, 200^2)$, correlation coeffecient $= 0.2$.

Typical simulation results are shown for each class of procedures. In Fig. 9.1(a)–
(e) five frames of a corrupted "Lena" image are shown. These frames were prepro-
cessed by a 5×5 GLS mask followed by a Wilcoxon robustized recursive least mean
square estimator. The results of the image reconstruction are shown in Fig. 9.1(g).

In the simulation of the missing value approach to the second stage of processing,
a 7×6 YS mask was used. Typical results are given for the "MIT building" image
in Figs. 9.2 and 9.3. It is obvious from the demonstrations, which are typical for
this class of problems, that this approach performs in a superior fashion if the
background noise is of the salt and pepper type.

9.6 Concluding remarks

This chapter addresses an important aspect of image processing, which is image restoration. Two restoration procedures based on a two-stage approach are introduced. For each class, the first stage consists of an edge detector based either on the GLS or YS masks. In the first class of procedures, the edge preserving image restoration is based on a recursive least square estimator. If the corrupting noise is of high variance, a robustized recursive version of the estimator is suggested. If the corrupting noise is of the salt and pepper type, the second stage of the restoration process is based on the missing value approach to the reconstruction process.

References

[1] E.L. Lehmann, *Testing Statistical Hypothesis*, John Wiley, New York, 1959.

[2] H. Scheffe, *The Analysis of Variance*, John Wiley, New York, 1959.

[3] G. Masson, *Statistical Design*, Prentice Hall, Englewood Cliffs, NJ, 1989.

[4] W.G. Cochran, G.M. Cox, *Experimental Design*, John Wiley, New York, 1957.

[5] L. Johnson, *Statistics and Experimental Design in Engineering and the Physical Sciences*, vol. 2, John Wiley, New York, 1977.

[6] R.G. Miller, *Simultaneous Statistical Inference*, McGraw-Hill, New York, 1967.

[7] M.G. Kendall, A. Stuart, *The Advanced Theory of Statistics*, Hafner, New York, 1963.

[8] J. Aron, L. Kurz, "A Statistical Approach to Edge Detection and Enhancement," *1973 International Symposium on Information Theory*, Ashkelon, Israel, June 1973.

[9] C. Mohwinkel, L. Kurz, "Computer Picture Processing and Enhancement by Localized Operations," *Computer Vision, Graphics and Image Processing*, vol. 5, pp. 401–424, 1976.

[10] D. Stern, L. Kurz, "Edge Detection in Correlated Noise Using Latin Squares Models," *Pattern Recognition*, vol. 21, no. 2, pp. 119–129, 1988.

[11] I. Kadar, L. Kurz, "A Class of Robust Edge Detectors Based on Latin Squares," *Pattern Recognition*, vol. 11, pp. 329–339, 1979.

[12] I. Kadar, L. Kurz, "Robustized Scalar Form of Gladyshev's Theorem with Applications to Nonlinear Systems," *Proceedings of the Annual Conference on Information Sciences and Systems*, Princeton University, pp. 297–302, March 1980.

[13] J.S. Huang, D.H. Tseng, "Statistical Theory of Edge Detection," *Computer Vision, Graphics and Image Processing*, vol. 43, pp. 337–346, 1988.

[14] D. Behar, *Masking Techniques for Line Detection and Edge Reconstruction in a Correlated Noise Environment*, Ph.D. dissertation, Polytechnic University, New York, 1988.

[15] R.B. Eberlein, "An Iterative Gradient Edge Detection Algorithm," *J. Computer Graphics and Image Processing*, vol. 5, no. 2, June 1976.

[16] R. Habestroh, *Four-Way Designs for Pattern Recognition in Digital Images,* Ph.D. dissertation, Polytechnic University, New York, 1987.

[17] A. Rosenfeld, A.C. Kak, *Digital Picture Processing*, vols. 1 and 2, Academic Press, New York, 1982.

[18] W.K. Pratt, *Digital Image Processing*, John Wiley, New York, 1978.

[19] A.K. Jain, *Fundamentals of Digital Image Processing*, Prentice-Hall, Englewood Cliffs, NJ, 1989.

[20] E. Chang, *Some Experimental Design Techniques in Image Processing*, Ph.D. dissertation, Polytechnic University, New York, 1981.

[21] E.H. Chang, L. Kurz, "Trajectory Detection and Experimental Designs," *Computer Vision, Graphics and Image Processing*, vol. 27, pp. 346–368, 1984.

[22] J. Aron, *Techniques of Classification and Feature Extraction: Analysis of Variance and Factor Analysis Based on Stochastic Approximation*, Ph.D. dissertation, New York University, New York, 1972.

[23] D. Bose, *Covariance Analysis and Related Techniques in Image Processing*, Ph.D. dissertation, Polytechnic Institute of New York, New York, 1982.

[24] S. Kariolis, *Some Statistical Methods of Image Processing and Reconstruction*, Ph.D. dissertation, Polytechnic Institute of New York, New York, 1979.

[25] M.K. Hu, "Visual Pattern Recognition by Moments Invariants," *IRE Transactions on Information Theory*, IT-8, pp. 179–187, 1962.

[26] G.B. Gurevich, *Foundations of the Theory of Algebraic Invariants*, Nordhoff, Groningen, The Netherlands, 1964.

[27] S.A. Dudani, K.J. Breeding, R.B. McGhee, "Aircraft Identification by Moments Invariants," *IEEE Transactions on Computers*, vol. C-26, no. 1, pp. 39–45, January 1977.

[28] F.W. Smith and M.H Wright, "Automatic Ship Photo Interpretation by the Method of Moments," *IEEE Transactions on Computers*, vol. C-20, pp. 1089–1094, 1971.

[29] Y-N. Hsu and H. Arsenault, "Rotation Invariant Digital Pattern Recognition Using Circular Harmonic Expansion," *Applied Optics*, vol. 21, p. 4012, 1982.

[30] Y-S. Abu Mostafa and D. Psaltis, "Recognition Aspects of Moment Invariants," *IEEE Transactions on Pattern Analysis and Machine Intelligence*, vol. PAMI-6, pp. 698–706, 1984.

[31] Y-S. Abu Mostafa and D. Psaltis, "Image Normalization by Complex Moments," *IEEE Transactions on Pattern Analysis and Machine Intelligence*, vol. PAMI-7, pp. 46–55, 1985.

[32] C.H. Teh and R. Chin, "On Image Analysis by the Methods of Moments," *IEEE Transactions on Pattern Analysis and Machine Intelligence*, vol. 10, no. 4, pp. 496–513, July 1988.

[33] M.O. Freeman and B.E. Saleh, "Moments Invariants in the Space and Frequency Domains," *J. Optical Society of America part A*, vol. 5, no. 7, p. 1073, July 1988.

[34] L. Jacobson and H. Wechsler, "A Theory for Invariant Object Recognition in the Frontoparallel Plane," *IEEE Transactions on Pattern Analysis and Machine Intelligence*, vol. PAMI-6, no. 3, pp. 325–331, May 1984.

[35] L. Massone, G. Sandini and V. Tagliasco, "Form-Invariant Topological Mapping Strategy for 2D Shape Recognition," *Computer Vision, Graphics Image Processing*, vol. 30, pp. 169–188, 1985.

[36] J. Neter, W. Wasserman, *Applied Linear Statistical Models*, Richard D. Irwin, 1974.

[37] S. Kariolis, L. Kurz, "Object Detection and Extraction: A Statistical Approach," *Proceedings of the Annual Conference on Information Sciences and Systems*, Princeton University, pp. 500–505, March 1980.

[38] E.H. Chang, L. Kurz, "Object Detection and Experimental Designs," *Computer Vision, Graphics and Image Processing*, vol. 40, pp. 147–168, 1987.

[39] I. Kadar, L. Kurz, "A Class of Three Dimensional Recursive Parallelpiped Masks," *Computer Vision, Graphics and Image Processing*, vol. 11, pp. 262–280, 1979.

[40] M.H. Benteftifa, L. Kurz, "An Approach to Object Detection in Correlated Noise," *23rd*

Annual Conference on Information Sciences and Systems, The John Hopkins University, March 1989.

[41] M.H. Benteftifa, L. Kurz "Two-Dimensional Object Detection in Correlated Noise," *Pattern Recognition*, vol. 28, pp. 755–773, 1991.

[42] M.H. Benteftifa, L. Kurz, "Rotation Invariant Object Detection Using ANOVA-Based Data Model," *Proceedings of the 28th Annual Allerton Conference on Communications, Control and Computers*, pp. 604–613, Allerton, IL, 1990.

[43] M.H. Benteftifa, *Object Detection Via Linear Contrast Techniquest*, Ph.D. dissertation, Polytechnic University, New York, 1990.

[44] M.R. Teague, "Image Analysis Via the General Theory of Moments," *J. Optical Society of America*, vol. 70, no. 8, pp. 920–930, August 1980.

[45] R.A. Messner and H.H. Szu, "An Image Processing Architecture for Real Time Generation of Scale and Rotation Invariant Patterns," *Computer Vision, Graphics and Image Processing*, vol. 31, pp. 50–66, 1985.

[46] R. Haralick, S. Shapiro, "Image Segmentation Survey," *Computer Vision, Graphics and Image Processing*, vol. 29, pp. 100–132, 1985.

[47] S. Horowitz, T. Pavlidis, "Picture Segmentation by a Tree Transversal Algorithm," *J. American Computing Machinery*, vol. 23, pp. 368–388, 1976.

[48] J. Browning, S. Tanimoto, "Split-and-Merge Image Segmentation Using a Tile by Tile Approach," *Pattern Recognition*, vol. 15, no. 1, pp. 1–10, 1982.

[49] R.H. Laprade, "Split-and-Merge Segmentation of Aerial Photographs," *Computer Vision, Graphics and Image Processing*, vol. 44, pp. 77–86, 1988.

[50] M. Pietikainen, A. Rosenfeld, I. Walter, "Split-and-link Algorithm for Image Segmentation," *Pattern Recognition*, vol. 15, no. 4, pp. 287–298, 1982.

[51] W. Frei, C. Chen, "Fast Boundary Detection: A Generalization and a New Algorithm," *IEEE Transactions on Computers*, vol. C-26, no. 10, pp. 988–998, October 1977.

[52] M.M. Trivedi, J.C. Bezdek, "Low-Level Segmentation of Aerial Images with Fuzzy Clustering," *IEEE Transactions on Systems, Man, and Cybernetics*, vol. SMC-16, no. 4, July/August 1986.

[53] S.K. Pal, "Segmentation Based on Measures of Contrast, Homogeneity, and Region Size," *IEEE Transactions on Systems, Man, and Cybernetics*, vol. T-SMC, pp. 857–868, September/October 1987.

[54] A. Perez, R. Gonzalez, "An Iterative Thresholding Algorithm for Image Segmentation," *IEEE Transactions on Pattern Analysis and Machine Intelligence*, vol. PAMI-9, no. 6, pp. 742–751, November 1987.

[55] R. Kohler, "A Segmentation System Based on Thresholding," *Computer Vision, Graphics and Image Processing*, vol. 15, pp. 319–338, 1981.

[56] J.S. Weszka, A. Rosenfeld, "Thresholding Evaluation Techniques," *IEEE Transactions on Systems, Man, and Cybernetics*, vol. SMC-8, no. 8, pp. 622–629, August 1978.

[57] J.S. Weszka, "A Survey of Threshold Selection Techniques," *Computer Vision, Graphics and Image Processing*, vol. 7, pp. 259–265, 1979.

[58] S.W. Zucker, "Region Growing: Childhood and Adolescence," *Computer Vision, Graphics and Image Processing*, vol. 5, pp. 382–399, 1976.

[59] J. Beaulieu, M. Goldberg, "Hierarchy in Picture Segmentation: a Stepwise Optimization Approach," *IEEE Transactions on Pattern Analysis and Machine Intelligence*, vol. PAMI-11, no. 2, pp. 150–163, February 1989.

[60] R.L. Kashyap, K. Eom, "Texture Boundary Detection Based on the Long Correlation

Model," *IEEE Transactions on Pattern Analysis and Machine Intelligence*, vol. PAMI-11, no. 1, pp. 58–67, January 1989.

[61] M. Unser, M. Eden, "Multiresolution Feature Extraction and Selection for Texture Segmentation," *IEEE Transactions on Pattern Analysis and Machine Intelligence*, vol. PAMI-11, no. 7, pp. 717–728, July 1989.

[62] G.R. Cross, A.K. Jain, "Markov Random Field Texture Models," *IEEE Transactions on Pattern Analysis and Machine Intelligence*, vol. PAMI-5, no. 1, pp. 25–39, January 1983.

[63] P.C. Chen, T. Pavlidis, "Image Segmentation as an Estimation Problem," *Computer Graphics and Image Processing*, vol. 12, pp. 153–172, 1980.

[64] S. Peleg, "Classification by Discrete Optimization," *Computer Graphics and Image Processing*, vol. 25, pp. 122–130, 1984.

[65] H. Derin, H. Elliot, R. Cristi, D. Geman, "Bayes Smoothing Algorithms for Segmentation of Binary Images Modeled by Markov Random Fields," *IEEE Transactions on Pattern Analysis and Machine Intelligence*, vol. PAMI-6, no. 6, pp. 707–720, November 1984.

[66] T. Kang, *Extraction of Primitive Features in Scene Analysis of Images Corrupted by Dependent Noise*, Ph.D. dissertation, Polytechnic University, New York, 1990.

[67] A.K. Jain, "Advances in Mathematical Models for Image Processing," *IEEE Proceedings*, vol. 69, no. 5, pp. 502–528, May 1981.

[68] C.W. Therrien, T. Quatieri, D.E. Dudgeon, "Statistical Model-Based Algorithms for Image Analysis," *IEEE Proceedings*, vol. 74, no. 4, pp. 532–551, April 1986.

[69] S.Y. Chen, W.C. Lin, C.T. Chien, "Split-and-Merge Image Segmentation Based on Localized Feature Analysis and Statistical Tests," *CVGIP: Graphical Models and Image Processing*, vol. 53, no. 5, pp. 457–475, September 1991.

[70] D. Stern, *Image Processing of Correlated Data*, Ph.D. dissertation, Polytechnic University, New York, 1987.

[71] M. H. Quenouille, *The Design and Analysis of Experiments*, Hafner, New York, 1953.

[72] L. Kurz, "Nonparametric Detectors Based on Partition Tests," in *Nonparametric Methods in Communications*, P. Kazakos and D. Kazakos, Eds. Marcel Dekker, New York, 1977.

[73] P. Kersten, L. Kurz, "Bivariate m-Interval Classifiers with Application to Edge Detection," *Information and Control*, vol. 34, pp. 152–168, June 1977.

[74] R. Dwyer, L. Kurz, "Sequential Partition Detectors with Dependent Sampling," *J. Cybernetics*, vol. 10, pp. 211–232, 1980.

[75] C. Tsai, L. Kurz, "An Adaptive Robustizing Approach to Kalman Filtering," *Automatica*, vol. 19, no. 3, pp. 279–288, 1983.

[76] C. Tsai, L. Kurz, "Robustized Maximum Entropy Approach to System Identification," *Proceedings of the International Federation on Information Processing and Systems*, September 1985.

[77] R. Dwyer, L. Kurz, "Characterizing Partition Detectors with Stationary and Quasi-Stationary Markov Dependent Data," *IEEE Transactions on Information Theory*, IT-32, pp. 471–482, July 1986.

[78] G.E.P. Box, "Non-Normality and Tests on Variance," *Biometrika*, vol. 40, pp. 318–335, 1953.

[79] G.E.P. Box, S.L. Anderson, "Permutation Theory in the Derivation of Robust Criteria and the Study of Departures from Assumption," *J. Royal Statistical Society*, B17, pp. 1–34, 1955.

[80] P. Kersten, L. Kurz, "Robustized Vector Robbins-Monro Algorithm with Applications to M-Interval Detection," *Information Sciences*, vol. 11, no. 2, pp. 121–140, 1976.

[81]   E.G. Gladyshev, "On Stochastic Approximation," *Theory of Probability and Its Applications*, vol. 10, no. 2, pp. 275–278, 1965.

[82]   G. Lomp, *Nonlinear Robust Detection and Estimation in Dependent Noise*, Ph.D. dissertation, Polytechnic University, New York, 1987.

[83]   J.M. Kowalski, *A Contribution to Robust Detection and Estimation in Dependent Noise*, Ph.D. dissertation, Polytechnic University, New York, 1992.

[84]   F.R. Hampel, E.M. Ronchetti, P.J. Rousseeuw, W.A. Stahel, *Robust Statistics*, John Wiley, New York, 1985.

[85]   P.J. Huber, *Robust Statistics*, John Wiley, New York, 1985.

[86]   M.T. Wasan, *Stochastic Approximation*, Cambridge University Press, New York, 1969.

[87]   A.E. Albert, L.A. Gardner, *Stochastic Approximation and Nonlinear Regression*, MIT Press, Boston, 1967.

[88]   H.B. Mann, D.R. Whitney, "On a Test of Whether One of Two Random Variables is Stochastically Larger than the Other," *Annals of Mathematical Statistics*, vol. 18, pp. 50–60, 1946.

[89]   E. Fix, L.J. Hodges, "Significance Probabilities of the Wilcoxon Test," *Annals of Mathematical Statistics*, vol. 26, pp. 301–312, 1955.

[90]   L.J. Hodges, E.L. Lehmann, "Estimates of Location Based on Rank Tests," *Annals of Mathematical Statistics*, vol. 34, pp. 307–317, 1963.

[91]   M. Chernoff, I.R. Savage, "Asymptotic Normality and Efficiency of Certain Nonparametric Test Statistics," *Annals of Mathematical Statistics*, vol. 35, pp. 102–121, 1964.

[92]   J. Sacks, "Asymptotic Distribution of Stochastic Approximation Procedures," *Annals of Mathematical Statistics*, vol. 29, pp. 590–599, 1958.

[93]   H. Robbins, S. Monro, "A Stochastic Approximation Method," *Annals of Mathematical Statistics*, vol. 22, pp. 400–407, 1951.

[94]   I. Kiefer, I. Wolfowitz, "Stochastic Estimation of the Maximum of a Regression Function,"*Annals of Mathematical Statistics*, vol. 23, pp. 462–66, 1952.

[95]   A. Dvoretzky, "On Stochastic Approximation," *Proceedings of the 3rd Berkeley Symposium on Mathematical Statistics and Probability*, Univ. of California at Berkeley, pp. 39–55, 1956.

[96]   S.A. Kesten, "Accelerated Stochastic Approximation," *Annals of Mathematical Statistics*, vol. 29, pp. 41–59, 1958.

[97]   J. Haberstroh, L. Kurz, "Line Detection in Noisy and Structured Background using Græco-Latin Squares," *CVGIP: Graphical Models and Image Processing*, vol. 55, no. 2, pp. 161–179, 1993.

[98]   M. H. Benteftifa, L. Kurz, "Feature Detection Via Linear Contrast Techniques," *Pattern Recognition*, vol. 26, no. 10, pp. 1487–1497, 1993.

[99]   Y. C. Trivedi, L. Kurz, "An Experimental Design Approach to Image Enhancement," *IEEE Transactions on Systems, Man, and Cybernetics*, vol. 22, no. 4, pp. 805–813, 1992.

[100] Y. C. Trivedi, L. Kurz, "Image Restoration Using Recursive Estimators," *IEEE Transactions on Systems, Man, and Cybernetics*, vol. 25, no. 11, pp. 1470–1482, 1995.

[101] A. Hussain, L. Kurz, "Robust m-Interval Detection Procedures for Strong Mixing Noise," *Information Sciences*, vol. 85, no. 1–3, pp. 113–125, 1995.

[102] G.R. Arce, N.C. Gallagher, T.A. Nodes, "Median Filters: Theory for one and Two Dimensional Filters," in T.S. Huang, ed., "Image Enhancement and Restoration," in *Advances in Computer Vision and Image Processing*, vol. 2, Jain Press, 1992.

Index

202